Antennas & Streaming

Reviews, comparisons, and step-by-step instructions

I0503967

By Ken Wickham

Antennas & Streaming
SBN-13: 978-1500399986
ISBN-10: 1500399981
Kindle ASIN: B00KWPCS9Y

First Edition : June 10, 2014

Updated First Edition : February 2, 2015

14 13 12 11 10 / 10 9 8 7 6 5 4 3 2 1

Dedication

This book is dedicated to all the antenna suppliers, video streamer devices, low priced subscription services, and free online media content providers

Thank You.

Ken

Forthcoming in this series

Internet + Wi-Fi phones

FTA Satellite

Sold Separately as Volume 1 and 2

Antennas + TV Program Guides
ISBN-13: 978-1499321135
ISBN-10: 1499321139
 Kindle ASIN: B00K1M63SK

Streaming Devices + Streaming Services
ISBN-13: 978-1500399276
ISBN-10: 1500399272
Kindle ASIN: B00KWO3P6K

Table of Contents

Sick of expensive unused channels

$157 monthly. $1884 annually.

That's how much money we were paying for cable television and internet. Sure we had 300+ channels. However, how many did we actually use?

How many of the paid channels do you use?

We watched two singing contest shows, a few suspense thrillers, a comedy sitcom, and a couple of sci-fi series. We actually only watched four TV shows at any given time, due to different show seasons. Occasionally, we rented a movie from the big red colored box outside store, or ordered through the mail some rental movie internet video from a monthly subscription. We did this because we didn't want to wait a year to watch these just-out-of-the-theater movies.

I noticed quickly, that the movies we received from the premium channels, we had already seen through the box, received via the mail or online streaming. There was only one premium TV show series that we watched 3 months every year.

How many TV shows come from network broadcast television?

Most of the TV shows we watched came from network broadcast television. Only three cable television series came from the part of cable that we paid. The portion that we paid for premium cable was about $100, only to watch three shows, and an occasional interesting history or biography documentary. That is $1,200 a year for three shows plus a few shows here and there.

Rise of the Cord Cutters

Sick of paying all this money for almost nothing, I had to think of an alternative. Many have cut their cable bills entirely off and become what many call "cord cutters".

Companies that have an eye towards the future are heading this wave of personalized-on-demand, subscription, or free commercial paid content.

Netflix, Google, Hulu, Amazon and even a few cable and satellite such as Xfinity, Dish Network, and DirecTV are companies are heading towards mobile, and more personalized content. TV Broadcasters are filling their websites with streaming episodes and content. A few live streams of news and a few networks already are on the net.

TV media devices are quickly becoming a part of everyday life

Remember antenna television

I remembered the time when all of my television was free coming from broadcast transmission towers received by a television antenna. We had rabbit ears sometimes modified with hangers and tinfoil, and sometimes one of those loop antennas for UHF.

After the digital conversion a few years ago, I was unsure if we could pick up any channels or how.

We watched several shows on the net anyway

Whenever we missed shows we realized that many networks would put some of their shows online, normally the last two to five episodes. We used this method to watch our favorite shows that we missed or hadn't seen because we were watching something else at that scheduled time. Normally the episodes appear on the network sites the next day.

Flash forward to where I am now

My mother and step-father are living north of Denver Colorado 50 miles. They have been spending $137 monthly for cable television. Analyzing what they actually watch, I told them I could save them maybe $80 at least a month and still receive most of the content that they watched every day. That's about $1000 a year in savings. They watch the news, health shows, reality contest shows, alien visitation documentaries, and ancient archeology documentaries and series. They also liked to listen to relaxation music.

I asked if they had an antenna.

My father-in-law said he did, and eventually found them in storage. I hooked them up one night and scanned for channels. Forty-two channels appeared.

2-1 KWGN - DT - The CW
2-2 KWGN – DT2 - This TV
4-1 KCNC – DT - CBS
5-1 KGWN – DT – CBS
5-2 KGWN – DT2 – Northern Colorado 5
5-3 KGWN – DT3 – The CW
6-1 KRMA-DT – PBSHD
6-2 KRMA - DT2 – V-ME
6-3 KRMA – DT3 - Create
14-1 KTFD - DT – Telefutura
14-2 KTFD – DT2 – Bounce
14-3 KTFD – DT3 - GetTV
20-1 KTVD – DT - MyNet

20-2 KTVD – DT2 MeTV

22-1 KFCT - DT - FOX

22-2 KFCT – DT2 – Antenna TV

24-1 CTVa – EWTN

24-2 TorahTV

24-3 Oorah

24-4 Impacto

27-1 KLWY – DT - FOX

27-2 KLWY – DT2 – ABC (KTWO rebroadcast)

31-1 KDVR – DT - FOX

31-1 KDVR – DT2 – Antenna TV

33-1 KQCK – DT – MundoFOX

36 Analog - Azteca

38-1 KPJR – DT - TBN

38-2 KPJR – DT2 – ChurchChannel

38-3 KPJR – DT3 – JuiceTV

38-4 KPJR – DT4 – TBNEnlace

38-5 KPJR – DT5 - Smile

44-1 KDNF – DT – Daystar

47-1 KRMA-DT – PBSHD

47-2 KRMA - DT2 – V-ME

47-3 KRMA – DT3 - Create

50-1 KCEC – DT - Univision

50-1 KCEC – LATV – LATV

59-1 KPXC - TV – ION

59-2 KPXC - TV2 - Qubo

59-3 KPXC – TV3 – IONLife

59-4 Shop

59-5 HSN

We started watching the over-the-air (OTA) antenna TV content for the next few days. After these few days of watching antenna television, my step-father made a call to his cable content provider to cut off his cable service entirely. I told him to keep the internet service. The Cable charged him large early disconnect fee, but we knew in the long run we would be saving money.

Apartment, indoor antenna

I live in an apartment currently 50 miles north of Denver. Because I live in an apartment, the rules prohibit attaching to the building. Indoor antennas are a necessity in this situation. My step-father had an amplified rabbit ears and convertor box already. What about places in the middle of nowhere?

In the middle of nowhere near Canadian border

I lived in Antler, North Dakota for a month last year. Although we received paid TV. I wondered what channels would even reach there, way out in the middle of nowhere. Here is what is listed. Of course, one would need an outdoor, rooftop, or tower antenna.

With a good roof antenna, my brother in Antler, ND, very near the US-Canada Border can pick up most major network broadcasts. CBS, NBC, FOX, ABC, PBS, World Channel, MN Channel, PBS Encore, Weather, Me-TV.

And with a repositioning north he might be able to pick up 3 Canadian channels which would give the two Canadian networks CTV and GTN.

Moderate Signal **KXMC** CBS 13-1 Hi-V

Channel	Aspect	Format	Programming
13.1	16:9	1080i	main KXMC-TV programming / CBS
13.3	4:3	480i	Weather

Moderate Signal **KMOT** NBC 10-1 Hi-V

Channels	Aspect	Format	Programming
10.1 / 8.1	16:9	1080i	Main KMOT/KUMV programming / NBC
10.2 / 8.2	4:3	480i	Me-TV

Weak Signal **KSRE** PBS 6-1 UHF

Channel	Video	Audio	Call Sign	Network/Programming		Nickname	Notes
06-1	40.3		1080i	DD2.0 PPB1	PBS	"Prairie Public Television"	
06-2	40.4		480i	DD2.0 PPB2	World Channel		
06-3	40.5		480i (w)	DD2.0 PPB3	MN Channel	"Minnesota Channel"	
06-4	40.6		480i	DD2.0 PPB4	PBS Encore	"Lifelong Learning"	

Weak Signal **KXND** FOX 24-1 UHF

Channel	Video	Audio	Call Sign	Network/Programming		Nickname	Notes
24-1	24.3		720p	DD5.1 KXND-DT	FOX	"Fox 24"	

Weak Signal **KMCY** ABC 14-1 UHF

Channel	Video	Audio	Call Sign	Network/Programming		Nickname	Notes
14-1	14.3		720p	DD2.0 KMCY-HD	ABC	"KMCY 14"	

And Possibly Canadian TV Channels

Very Weak Signal CKYB CTV 4, CKND GTN 2, CIEW CTV 7

Summary of Antenna Setup Instructions

Before you are able to receive and use over-the-air broadcast signals, you must take a few steps to plan, purchase the needed equipment and accessories, set everything up, and set up useful TV program guides.

These steps begin with analyzing your TV which leads you to three tools that can help you figure out the possible over-the-air content you might be able to receive. A list of possible antenna and accessory retailers and online stores is given. The book then gives some universal steps to setting up your channels.

1. [   ] If you don't have an antenna and only a TV first write down the qualities of your TV or computer monitor you will be watching your TV. This information includes connector types, TV tuner, and screen resolution. Record this info on the planning information sheet in section I. You may print off a copy of the planning pages.

2. [   ] Next, use the online tools to figure out signal tower directions, distances, and signal strength. For more information how to use these tools read the *TV Transmission Tower Information* chapter. Record this info on your planning pages in the appropriate section in section II

3. [   ] Also, complete the Channel Information sections to figure out the available channels, networks, and sub channels using the available tools explained also in that same chapter. If two or more of the same channel are broadcast from different towers, there is room enough to record the two closest towers. Record this information in section III in the planning pages.

4. [   ] With this information you should have a good idea what content is available over-the-air. You will also be equipped with the information needed to purchase an antenna.

5. [ ] Using the information compiled and written down from step 2 and knowledge of possible signal blocking obstacles, you can purchase the appropriate antenna. Use the information in *What Kind of Antenna Do I Need* chapter.

6. [ ] Scout for retail antennas, online antennas, or professional installers. See the list of possible places that sell antennas at the back of the book. Consult a local antenna installer if you need help installing a roof antenna. Learn the return policies just in case things don't work out as planned. Purchase the best antenna for your planned situation within your budget. Keep all receipts in case things don't work out as planned.

7. [ ] Install or have someone with the required skills install your antenna. Indoor antennas should be easy enough to do on your own or by someone with basic antenna knowledge. The antenna coaxial cable should go in the antenna input of the TV tuner. If you have an old TV, converter boxes then have an output which hook up to the TV.

8. [ ] Set up your TV to scan for channels.

9. [ ] Set up a *TV Program Guide* to help you plan upcoming content.

10. Enjoy!

Planning your TV entertainment services

TV Antenna

In the book or on separate pieces of paper, please write down the following information in order to help you setup an alternative to expensive paid TV

I. What type of television connectors does your TV have? How many?

- Coaxial _____
- HDMI _____
- S-Video _____
- Compodite (LR(red white) audio, (yellow) video _____
- Component (Green - Y, blue – Pb, red – Pr; all three are video, audio is separate)
- USB _____
- VGA _____
- DVI _____

What kind of TV tuner will you be using? (Check or write down each one)
   A. Built in TV   [ ]
   B. TV converter box   [ ]
   C. Computer TV tuner   [ ]

What is your TV resolution?
   A. 480i (analog) [ ]
   B. 480p (digital) [ ]
   C. 720p [ ]
   D. 1080p [ ]
   E. WQHD 1440p [ ](Highest currently possible)

II. What general direction are the transmission towers? [FCC Method] (Write Direction)

- Strong signals (Green) _____

   _____

- Moderate signals (Yellow) _____

   _____

- Weak signals (Orange) _____
- No signals (Red) _____

How many stations are: Low-V(HF)_____, Hi- V(HF)_____, UHF_____?

What is the exact direction and distance to the towers? [TVFool.com Method]

Channel Real - Virtual - Network - Distance - Direction - True - Magnetic

_____ _____ _____ _____ _____ _____ _____

_____ _____ _____ _____ _____ _____ _____

Channel Real	Virtual	Network	Distance	Direction	True	Magnetic

III. What Networks are listed? [Use Rabbitears.com Method] (check or write down each one)

Networks

CBS [ ] Chan _____Dir_____Dist_____ || Chan _____Dir_____Dist_____

ABC [ ] Chan _____Dir_____Dist_____ || Chan _____Dir_____Dist_____

NBC [ ] Chan _____Dir_____Dist_____ || Chan _____Dir_____Dist_____

Fox [ ] Chan _____Dir_____Dist_____ || Chan _____Dir_____Dist_____

CW [ ] Chan _____Dir_____Dist_____ || Chan _____Dir_____Dist_____

ION [ ] Chan _____Dir_____Dist_____ || Chan _____Dir_____Dist_____

Qubo [ ] Chan _____Dir_____Dist_____ || Chan _____Dir_____Dist_____

IONLife [ ] Chan _____Dir_____Dist_____ || Chan _____Dir_____Dist_____

Public

PBS [ ] Chan _____Dir_____Dist_____ || Chan _____Dir_____Dist_____

Create [ ] Chan _____Dir_____Dist_____ || Chan _____Dir_____Dist_____

World [ ] Chan _____Dir_____Dist_____ || Chan _____Dir_____Dist_____

MHZ Worldview [ ] Chan _____Dir_____Dist_____ || Chan _____Dir_____Dist_____

NHK World (Japan) [ ] Chan _____Dir_____Dist_____ || Chan _____Dir_____Dist_____

France24 (French, English, Arabic) [ ] Chan _____Dir_____Dist_____

V-Me (Spanish) [ ] Chan _____Dir_____Dist_____ || Chan _____Dir_____Dist_____

Shows and Movies

ThisTV [ ] Chan _____Dir_____Dist_____ || Chan _____Dir_____Dist_____

GetTV [ ] Chan _____Dir_____Dist_____ || Chan _____Dir_____Dist_____

Antenna I V [ ] Chan _____Dir_____Dist_____ || Chan _____Dir_____Dist_____

MeTV [ ] Chan _____Dir_____Dist_____ || Chan _____Dir_____Dist_____

MyNetwork [ ] Chan _____Dir_____Dist_____ || Chan _____Dir_____Dist_____

BounceTV [ ] Chan _____Dir_____Dist_____ || Chan _____Dir_____Dist_____

CoziTV [ ] Chan _____Dir_____Dist_____ || Chan _____Dir_____Dist_____

RetroTV [ ] Chan _____Dir_____Dist_____ || Chan _____Dir_____Dist_____

Movies! [ ] Chan _____Dir_____Dist_____ || Chan _____Dir_____Dist_____

Music

Zuus Country [ ] Chan _____Dir_____Dist_____ || Chan _____Dir_____Dist_____

CoolTV [ ] Chan _____Dir_____Dist_____ || Chan _____Dir_____Dist_____

Mens

Tuff TV [ ] Chan _____Dir_____Dist_____ || Chan _____Dir_____Dist_____

Weather

Accuweather [ ] Chan _____Dir_____Dist_____ || Chan _____Dir_____Dist_____

WeatherNation [ ] Chan _____Dir_____Dist_____ || Chan _____Dir_____Dist_____

Local Weather [ ] Chan _____Dir_____Dist_____ || Chan _____Dir_____Dist_____

American One [ ] Chan _____Dir_____Dist_____ || Chan _____Dir_____Dist_____

TV Guide

TV Scout [ ] Chan _____Dir_____Dist_____ || Chan _____Dir_____Dist_____

Shopping

QVC [ ] Chan _____Dir_____Dist_____ || Chan _____Dir_____Dist_____

HSN [ ] Chan _____Dir_____Dist_____ || Chan _____Dir_____Dist_____

JewelryTV [ ] Chan _____Dir_____Dist_____ || Chan _____Dir_____Dist_____

Religious

TBN [ ] Chan _____Dir_____Dist_____ || Chan _____Dir_____Dist_____

Daystar [ ] Chan _____Dir_____Dist_____ || Chan _____Dir_____Dist_____

Enlace (Spanish) [ ] Chan _____Dir_____Dist_____ || Chan _____Dir_____Dist_____

Smile of a Child [ ] Chan _____Dir_____Dist_____ || Chan _____Dir_____Dist_____

JuiceTV [ ] Chan _____Dir_____Dist_____ || Chan _____Dir_____Dist_____

The Church Channel [ ] Chan _____Dir_____Dist_____ || Chan _____Dir_____Dist_____

3ABN [ ] Chan _____Dir_____Dist_____ || Chan _____Dir_____Dist_____

Cornerstone [ ] Chan _____Dir_____Dist_____ || Chan _____Dir_____Dist_____

EWTN [ ] Chan _____Dir_____Dist_____ || Chan _____Dir_____Dist_____

FamilyNet [ ] Chan _____Dir_____Dist_____ || Chan _____Dir_____Dist_____

Spanish

Univision [ ] Chan _____Dir_____Dist_____ || Chan _____Dir_____Dist_____

Telemundo [ ] Chan _____Dir_____Dist_____ || Chan _____Dir_____Dist_____

MundoFox [ ] Chan _____Dir_____Dist_____ || Chan _____Dir_____Dist_____

UniMas [ ] Chan _____Dir_____Dist_____ || Chan _____Dir_____Dist_____

Estrella TV [ ] Chan _____Dir_____Dist_____ || Chan _____Dir_____Dist_____

Azteca America [ ] Chan _____Dir_____Dist_____ || Chan _____Dir_____Dist_____

LATV [ ] Chan _____Dir_____Dist_____ || Chan _____Dir_____Dist_____

Existos TV [ ] Chan _____Dir_____Dist_____ || Chan _____Dir_____Dist_____

Zuus Latino [ ] Chan _____Dir_____Dist_____ || Chan _____Dir_____Dist_____

Health and Wealth

Living Well Network [ ] Chan _____Dir_____Dist_____ || Chan _____Dir_____Dist_____

Biz Television [ ] Chan _____Dir_____Dist_____ || Chan _____Dir_____Dist_____

What type of antenna do you have (will you get)? (circle which one is best)

Indoor unamplified (unblocked, less than 25 miles from transmission tower): Urban and Suburban. Forms- flat, rabbit ears, and loop. Great value if unblocked by trees, buildings, and near towers.

Indoor amplified (booster) (unblocked, less than 40 miles from transmission tower): Near rural areas, suburban, urban. Forms- similar to unamplified except with a plug in for the amplifier.

Attic amplified (less than 45 to 50 miles from transmission towers): Form basically a small enough outdoor antenna, instead mounted on a poll or pillar in the attic. You may need to construct some basic mount or pole depending on your attic.

Outdoor amplified (unblocked, normally less than 70 miles from transmission tower) Rural areas, near rural areas, suburban, urban. These can be

Breadth and depth (circle which one is best)

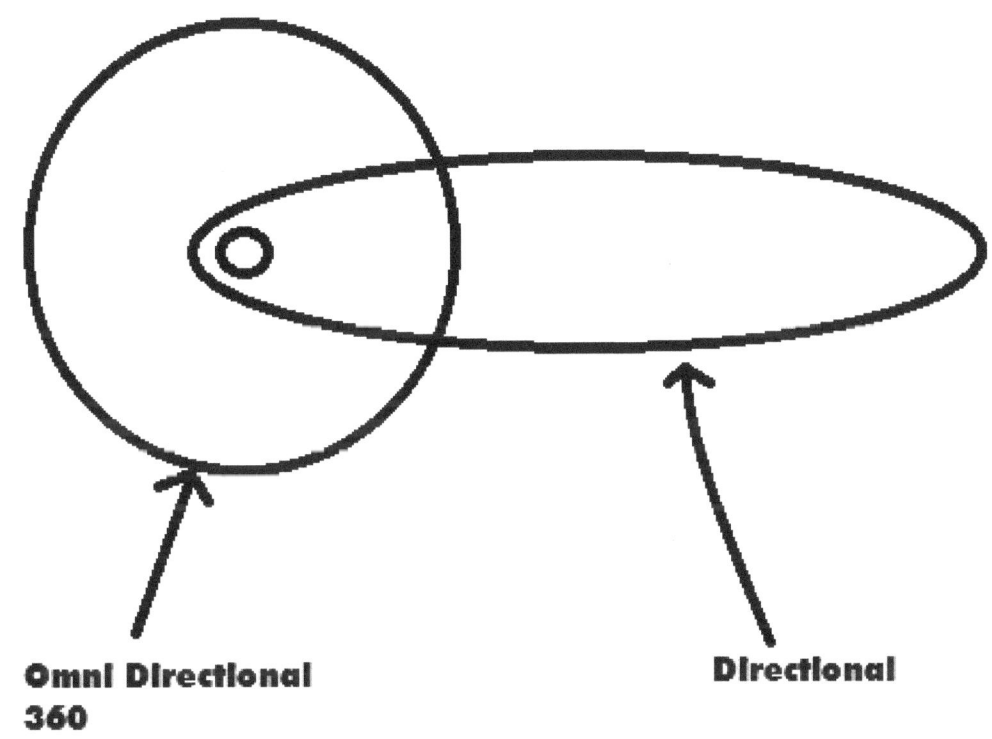

Omni Directional 360

Directional

General arc of two major types of TV antennas
Ranges are not drawn to scale

Drawing for illustrative purposes only. The range and shape are not entirely accurate. See professional specifications for actual shape and range of each antenna and type. By Ken Wickham

Omnidirectional (360degrees): In the center of multiple towers surrounding under 40 mile range.

Multi-directional: an antenna that picks up signals for a wider angle, 50 to 90 degrees, though it loses some distance vs directional. Distance however is further than 360 degree antennas

Directional fixed: Signals coming from one direct, within (20 degrees best, 35 degree limit of each other).

Rotatable (rotor): Benefits of directional antenna range, though you can rotate the antenna to aim at different directions. Signals coming from all around you. Less than 70 miles from transmission.

Pick a TV Program Guide

[ ] The built-in TV EPG guide is sufficient.

[ ] TVListings.AOL.com

[ ] TitanTV.com

[ ] TVGuide.com

[ ] Zap2lt.com

[ ] TV.com

The first thing I bought for my first antenna setup in Mississippi was a nice amplified indoor antenna. We lived near a medium populated city, and did not need a roof or attic antenna. I chose a *Micron-R ClearStream Micron Indoor UHF DTV Antenna with Reflector Screen*. After hooking the antenna on, plugging in the amplification power adaptor, and doing a channel antenna search, only a few fuzzy channels appeared. I realized this was an older television needing an analog to digital convertor box. My mother-in-law used one of these boxes connected to her roof antenna to get television out in the farming country region. We bought a TV signal analog to digital converter box, hooked it up, and tried a channel search. The specs showed only a 35 mile range, though we pulled in signals from 40 miles away with a little adjustment of the direction and adding a booster.

What Kind of Viewing TV or Computer Tuner do you own?

The first step you need to do figure out if your TV has a built in digital TV tuner, or is an older analog TV tuner. If you are going to use a computer as a TV you will need to check if you have a TV tuner card (normally indicated by a coaxial input port on your computer.

Built into TV [ ]
Converter Box [ ]
Tuner Card [ ]

At the same time, you may want to make a list of all the devices you plan on including in your entertainment system.

Accessories
Antenna [ ]
Media Player [ ]
Media Streamer [ ]
DVR [ ]
DVD player [ ]
Blue-ray player [ ]
Video game console [ ]
Sound system or speakers [ ]

TV Connectors
Next, you need to know what type of connectors are included on your television(s) you will be using.

HDMI

HDMI emerging in 2004, HDMI has replaced coaxial as the high definition connection of modern televisions. 90% of HDTVs by 2007 had HDMI connectors. By 2009, all digital televisions had at least one HDMI connectors. Many converters exist to convert different type of connectors and wires into HDMI. The signal probably will not play at the highest resolution however. Be cautious of having too many connectors, splitters, and wires which leads to signal loss and overall masses of wire octopuses.

S-Video

S video has a max 480i/576i signal definition. S-Video is slightly better than composite video, using 2 channel encryption instead of one. You may have to use this for older TVs though composite a/v is normally more common.

Composite A/V

Older TVs might need to use the composite port. Composite has a max of 480i/576i. This is the 3 color wires of yellow for video, and red and white for audio. Composite A/V is slightly lower quality compared to S-Video because it only uses one channel instead of the two that S-Video uses.

Ethernet

Cat-5e is the current standard for Ethernet wiring. Ethernet wiring is mainly used to connect devices by wire to routers and modems.

Coaxial

Older still is the basic coaxial input and output. RG-6 is the current standard for coaxial cable.

USB

USB port on TVs can run movies, listen to music, and look at pictures from a thumb drives. It can also update the TVs firmware by downloading the software putting it on a thumb drive then inserting the drive into the port. External hard drives can also be used to play content connecting directly to the TV.

TV resolution, aspect ratio, formats

When the entire United States switched over from analog to digital service, the entire country was thrown into a little frenzy. What emerged however has its advantages and disadvantages.

The old analog signal was a continuous updating signal. The further and weaker the signal, you might still be able to get a fuzzy picture with some sound. The digital signal is more of an all-or-nothing type signal. This means you either get it or don't. Although in reality, you can get some channels that are pixelated. Normally those pixelated channels you can tune in with proper repositioning of the antenna, change of height, or addition of a booster (amplifier).

Modern signal types

Broadcaster digital terrestrial television (DTV) broadcasting. Broadcasters normally transmit one or both types of picture formats, which vary in size and aspect-ratio. High definition television (HDTV) for the transmission of high-definition video and standard-definition television (SDTV).

The i and p in the resolution

In the signal transmission the gap is called *interlacing video* (symbol = i) in the analog method and progressive (symbol = p). In interlacing they sent half a frame at a time, first the odd lines then the even lines. Now they send line-by-line. So when you read 720p, this actually means 1280x720 digital *progressive scan* signal in that it receives a signal line by line. This reduced eyestrain from interline twitter making images and movement more smooth.

Old analog signals (for comparison)

CGA computer monitors had a resolution of 320x200 up to 640x200 4 bit 16 color.

VGA computer monitors (1987) introduced 640x480 up to 800x480 16 bit in 4:3 aspect ratio. VGA has 256 colors in 320x200 mode. The second resolution is the VHS and Betamax resolution. 480i is also the broadcast resolution of old analog televisions.

Current HDTV signals

HD 480p has 640x480 pixel or 4:3 aspect ratio (4 units wide by 3 units high) is SDTV. This is closest to DVD quality which is 720x480 which can be shrunk to fit SDTV, or play in widescreen mode with dark space above and below the picture.

HD 720p widens the resolution to 1280x720 16:9 aspect ratio.

HD 1080p resolution of 1920 × 1080 at a 60 Hz with 16:9 aspect ratio. This is Blue-ray resolution.

What Kind of Antenna Do I Need

With the information gathered from primarily TVFool.com and the FCC, you can begin to process of choosing an antenna. Special focus is given to your range, the signal RF channels, and the compass direction of the towers.

Estimating antenna range by type

A rule-of-thumb general estimate of range of TV Antenna is 70 miles for **_outdoor antennas_** with ideal conditions due to the curvature of the earth. Using an indoor antenna generally cuts the signal in half, so 35 miles is a general range rule for **_indoor antennas_**. Some VHF signals with proper conditions may pick up signals 100 miles away. UHF (majority of signals) have a much shorter range. A basic rabbit ears **_unamplified_** is good for about 10 miles.

What type of signal frequencies are you trying to receive (UHF v VHF or UHF&VHF)?

The FCC, TVFool.com, and Rabbitears.com tell the RF, Physical, or true channel that the signal transmits using. Also they tell the Virtual, Display channel. Use the RF channel to figure out how many real VHF channels you have. Most channels will be UHF.

VHF

VHF covers the frequencies channel range of 2 to 13. Channels 2 through 6 are Low Band VHF channels. Channels 7 to 13 are High Band VHF channels.

UHF

UHF covers the frequency channels 14 to 51. UHF frequencies advantage is that there normally isn't as many interfering signals due to devices, computers, and other noise.

Will you focus the signal on one direction or need multi direction or all-around signal reception?

Directional – Benefit is longer range. Disadvantage is narrower width of signal reception.

Fixed – Antenna is facing one direction. Advantage is if all signals come from one direction and have a narrow width.

Rotatable – Antenna can move around. You gain the advantage of having a directional long range antenna. However, you can rotate the antenna to pick up multiple directions. Thus you create a 360 degree long range antenna ability.

Multi-Directional – Means somehow it can receive signals from more than one direction. Depending on the type of multi-directional it may or may not receive signals 360 degrees.

Omnidirectional – You can receive signals from any direction barring obstacles. Disadvantage is you will have a shorter range. This type of antenna is great if you are surrounded by many signal towers that are not that far away like when you are in a city.

Will you be able to place the antenna outdoor, in the attic, or will it be indoor room only?

Most of the signal tools in this book will list signal strength by colors green representing indoor antenna, yellow represents attic antenna, and orange represents outdoor antenna needed.

Do you need any amplification (booster)?

In my experience an amplification can help gain channels that you can only partially or barely receive. You will not gain a ton of channels, rather you might be able to pick up a few don't come in clear. Amplifiers also amplify noise, which could make some channels worse. Too much gain causes overload.

You should get an antenna with the accessories that give you a little more gain than you need. Excessive gain leads to overload. Too little gain will lead to inability to receive the enough signal.

Long coaxial cable runs might require a pre amp or amplifier. Use of splitters might require use of amplifiers.

TV signal reception

The old television analog needed less of a signal strength for reception than digital. In addition, in analog, the signal frequency identified the channel. Most channels now broadcast on Ultra High Frequency (UHF) band which are line of sight (LOS) and do not go around or through signals. They may or may not be identified by their radio frequency (RF channel). Some are identified by their DTV or virtual channel designation rather than their actual RF channel. When scanning for channels, you will see the channels accepted using the RF channel number. When you actually change the channel using the remote, it will show up using the DTV channel number, if different from the actual RF channel.

What do I need to look for in a TV tuner?

ATSC

ATSC is the signal standard for North America which replaced the most of the old analog signal standard NTSC in the US on June 12, 2009. ATSC means Advanced Television Systems Committee.

Note on other signal types standards

You will probably come across several different standards for signal transmission.

NTSC

NTSC is the old signal standard and signal standard for analog TV used primarily before June 12, 2009. A few analog channels still exist. NTSC means National Television System Committee.

QAM (DVB-C)

QAM is the digital cable standard. QAM means quadrature amplitude modulation. The signal is called DVB-C, also meaning Digital Video Broadcasting — Cable.

DVB-S is the digital satellite standard. DVB means Digital Video Broadcasting — Satellite.

What about older TVs?

TVs made before accepting the new ATSC TV standard will require a TV *converter box*. The converter box, instead of the set channel changer, will control the changing of the channels. The set is normally set on 3, 4, or some sort of input source mode. Read the instructions of the converter box for details. The converter box wired between the antenna (coaxial IN) and the TV (coaxial, HDMI, or component out).

Normally a wireless remote controls the channels and volume. Programmable remotes can replace lost or damaged remotes, or consolidate several remotes for several devices.

What do I need to turn my computer or laptop into a TV?

Computers require a TV tuner in order to receive TV signals, also some sort of software is needed to display the TV channels. Look for a TV tuner that can receive ATSC signals. Also you will need the drivers, which come on a DVD disk. Modern windows can automatically searches for the drivers, if you are connected to the internet. In Microsoft Windows you can use Windows Media Center to play live TV. Mac uses Front Row as its Media Center main software.

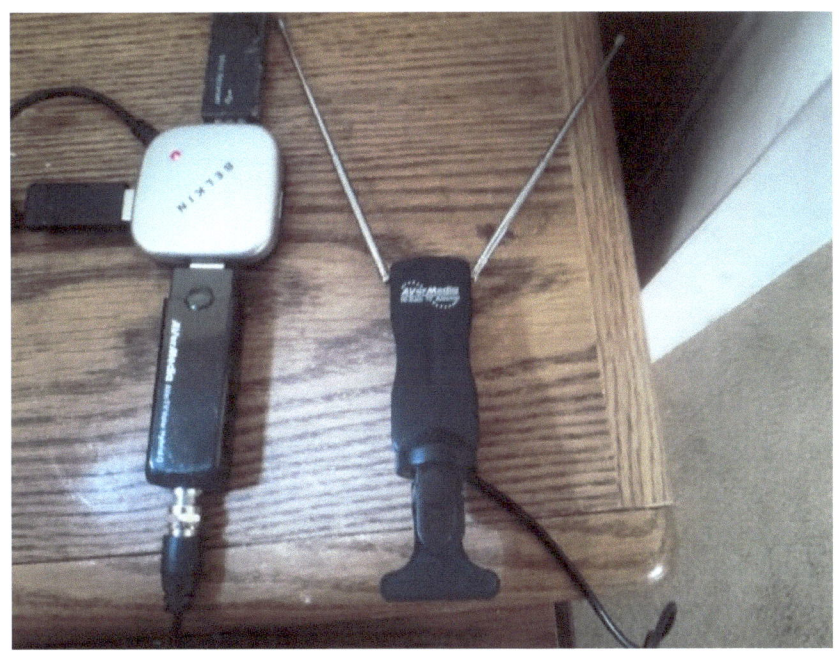

My computer TV tuner

TV Antenna types

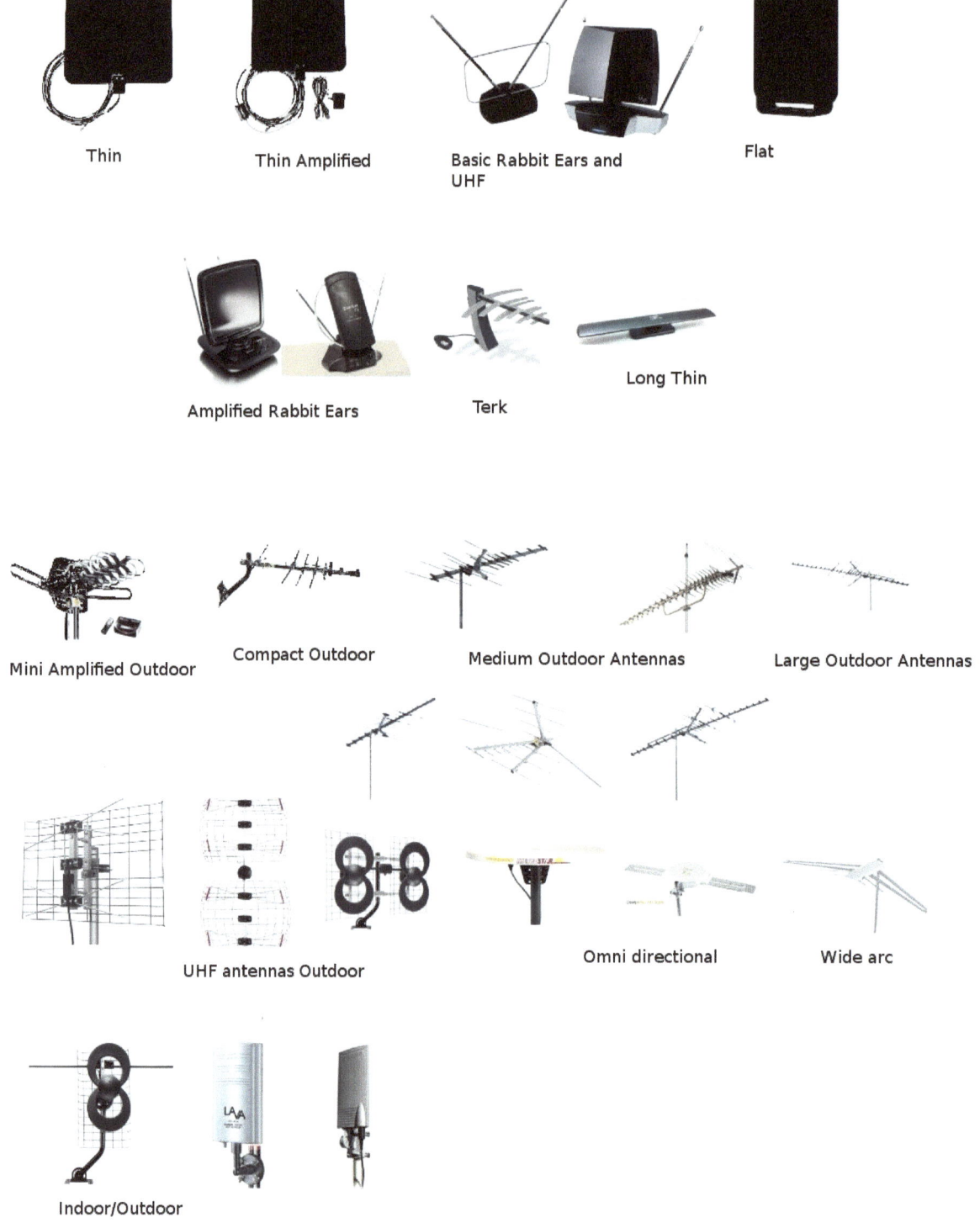

This chart for comparison purpose shows the variety of types one can make depending on the type of TV antenna needed.

Making a TV Antenna

We following these instructions and made a great antenna higher gain than the amplified indoor antenna that we bought.

"How to get HD channels with a $2 homemade antenna" by John Matarese posted 9:16 AM, Feb 11, 2014. http://www.abcactionnews.com/money/consumer/dont-waste-your-money/how-to-get-hd-channels-with-a-2-homemade-antenna

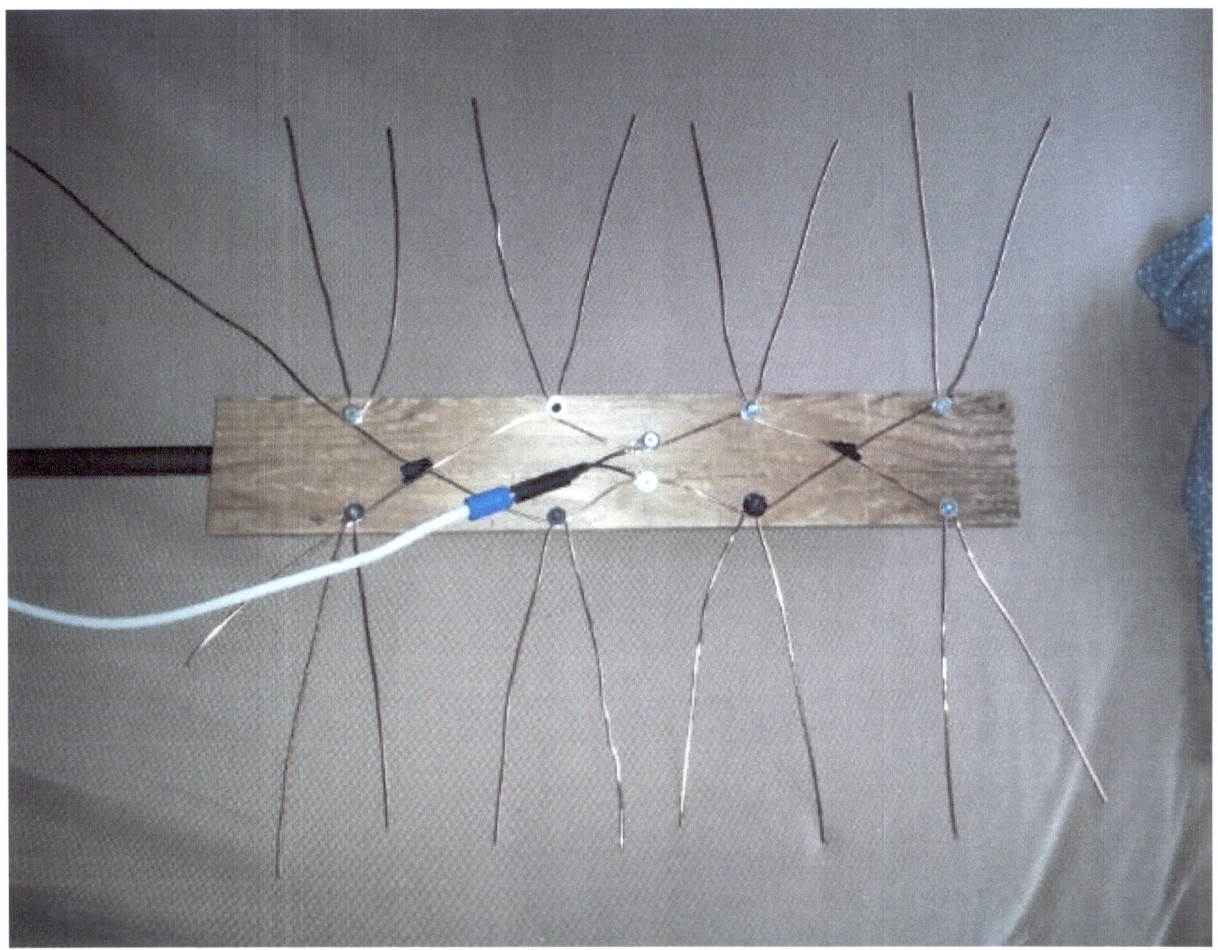

Notes on TV accessories

Wire

Coaxial cable – the wire that connects antennas to the TV. For TV antennas, RG-6 is the common wire type. RG- 59 is for satellite systems. The longer the coaxial runs, the weaker the signal. Short coaxial runs with no or few connections or splits would result in the best signal connection.

Color of wires

White – Normally used indoor along trims, since most indoor rooms are white or very light.

Black – Normally used in walls and outdoors.

75 ohm coaxial RG – 6 is the standard for TV antennas that replaced the old 300 ohm twin lead (flat wide looking wire). Older antennas on older homes may still have the old wire. If so attach a 300 ohm to 75 ohm Matching Antenna Balun Transformer and then run a new line of coaxial RG – 6

Solid wire cores

      Good – Standard coaxial - cable copper coated steel wire.

      Better – Solid copper coaxial cable (though more expensive)

Shielding

Shielded or regular – Shielded is normally indicated by Quad-shielding. Regular only has one or two shields

Other Ratings

Flooded/Direct Burial – This coaxial can be buried under ground in conduit. Flooded has a sticky outer coating to help protect it from the moisture.

Outdoor cable also have a UV rating.

UL Ratings

CMP – Plenum – installing in ducts, fire resistant, low smoke and toxin producing.

CMR – Riser – used to prevent spread of fire in elevator shafts from floor to floor.

CM – Used to prevent fire in areas other than plenum or risers.

Coaxial **F-type connectors**

      Worst - Twist on

      Better – Crimp on. Look for outdoor water resistant with O-rings and sealing gel

      Best – Compression fittings

Ground – Outdoor antenna masts need to be grounded with #8 or #10 copper or aluminum wire directly to a copper coated steel ground rod 3 feet deep at least. The wire should be attached to the mast directly good and solid, scrapping off paint if necessary (then painting or sealing over the connected ground). There should be no more than 90 degree turns and turns should be gradual and smooth. Do not attach to pipes or plumbing.

Splitters & Combiners

Splitter – Takes one input and divides the signal into two or more outputs. The higher the quality, the less the signal loss. All splitters divide signals weakening the signal strength. The less the number of divisions and connections the better the signal.

High Bandwidth Splitter – Used to split higher bandwidth of HDTV like 1GHz or 2GHz

Combiner – Takes two input sources (ex 2 antennas) and combines the signals into one output.

Splitter/Combiners – These can be used as either a splitter or a combiner.

Amplifiers – Amplifiers come in two major types. Pre Amps and Signal Amplifiers
Pre Amp - used for coaxial runs longer than 50 feet or when towers are 45 miles or greater distance.

Signal Amplifier – Amplifiers can be used sometimes integrated with combiners and splitters. Or it can be used to recover lost signal after a splitter.

Outdoor antenna mount types

Tripod (roof peaks)
Eave mount kit
Chimney
Wall
J-Pole

When mounting on a roof or wall, ensure that

Rotors – Rotators – Mast mounted motor allowing the antenna to orient towards different directions via an indoor control.

Don't place outdoor antennas near power lines.

Don't bend coaxial too sharply

Apply silicon grease or acrylic insulator to older antennas connections with a Balun transformer.

Four Free Online Tools to Get Antenna Range, Distance, Sub channel, Network Information

A) TV Transmission Tower Information

What general direction are the transmission towers? (Easy Quick Method)

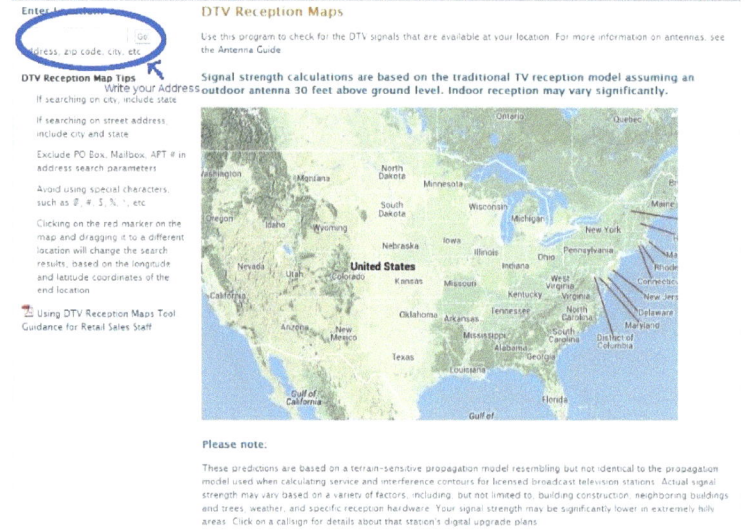

Where to type your address.

FCC Digital TV Maps

http://transition.fcc.gov/mb/engineering/dtvmaps/

1. Enter your address into the location box circled in blue.
2. After entering your address **enter** or click on the **GO** button.
3. Signal strength is show primarily color coded.
   a. The call signs shaded in green you should easily be able to pick up with an indoor antenna, unless some obstacle impedes your signal such as a hill or large building. Amplification or booster may increase some channel reception.
   b. Yellow means you may need an attic antenna.
   c. Orange you need an outdoor roof antenna.
4. The Network is the primary station network. Other networks and digital subchannels can be contained within each channel broadcast which you will find using the Rabbitears tool.
5. Note the band. The band helps you to determine which type of antenna you need. How many Low-V, Hi-V, UHF

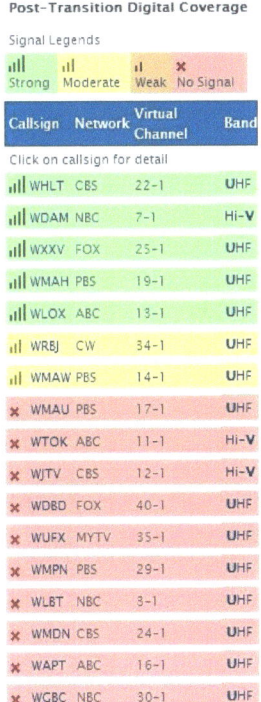

A quick FCC check on channels shows this for Hattiesburg, MS. This is all the major networks

6. Click on each callsign to see expanded information for that tower. Additional information show will be as followed.
   a. A location on the map will be show. You can zoom in and out to better see the location of the TV tower transmitter.
   b. The virtual channel number and the RF channel broadcasted.
   c. The receiving power in dBm. The combination of antenna, and amplifiers, minus the cord length, distance loss, obstacle interference, splitter loss must equal above zero to be able to view the channel.
   d. A compass direction and general direction. In addition the location will be shown on the map.
   e. You can click on the **Gain/Loss Map** link to see a map of the viewing area.
7. For now just write down the call signs and general direction for stations that come up green, yellow, orange, and red. Your main focus now will be to figure out which directions are most important. Are there any repeat stations in different directions? If so, the one in the dominant direction might be more important and easier to receive.

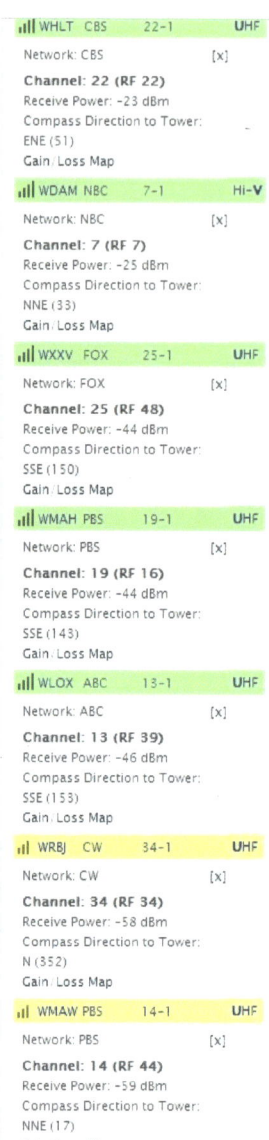

Checking the map of the main signals shows the directions of the transmission towers. As you can see most of the channels are either NE, NNE or SSE. I see where these transmission towers are by clicking on each of the network call letters. So these are the main direction focus.

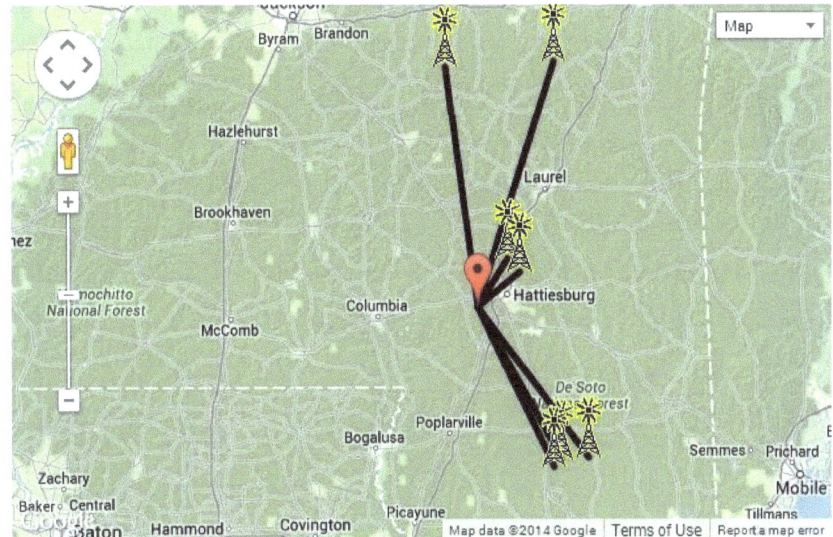

The channels on the list above come from the 60 mile range. Only two transmission towers are 14 miles away; NBC and CBS. Three are 40 miles away; PBS, FOX, and ABC. Two are 60 miles away; CW and another PBS. You can see on the map above the direction of these towers, I get the exact distance from another tool that I will mention from TVFool.com.

No Signal Transmission Towers

Just to know if I get adventurous, where are these red channels coming from. I click on these to find out, while deselecting the ones I probably will get. Mostly NW and W are the channels that would be less likely to reach my house in Hattiesburg. One comes from the NNE. These channels are about 90 miles away according to TVFool.com.

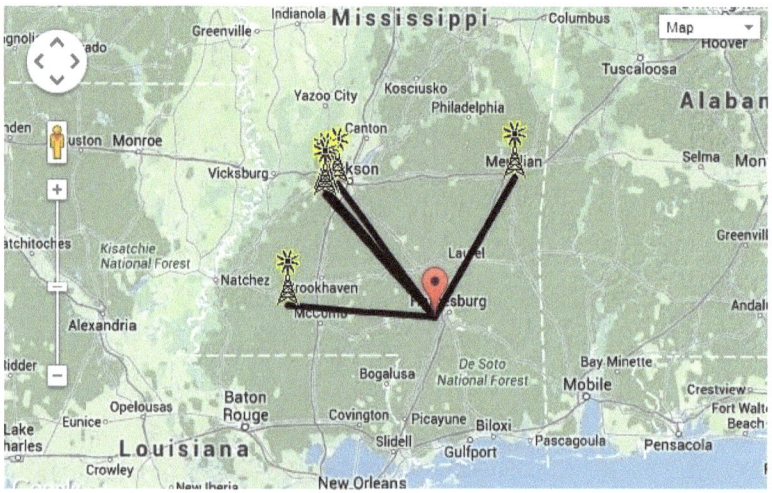

Another Example

For a quick peak you can go to http://transition.fcc.gov/mb/engineering/dtvmaps/ and enter your address. You may or may not be able to pick up the channels on this fast list. This list assumes you have a 30 foot high antenna without any obstacles. If you have an indoor antenna or have buildings and hills in the way, you may not get some of these channels. You can click on the station callsign on this list to see which direction the signal is originating from.

One of the signal towers impeded by buildings.

Example: When trying it for my area it shows 20 call signs or networks. 6 of them, however we do not pick up. Further analysis shows that these channels originate from a direction where 4 huge buildings block. Also important to note, 3 of these channels are repeats of the same network from different antennas. So of the 20 possible channels we can pick up, 6 are blocked by buildings, and three are repeated twice. So in total 11 unique network call signs are listed in this list.

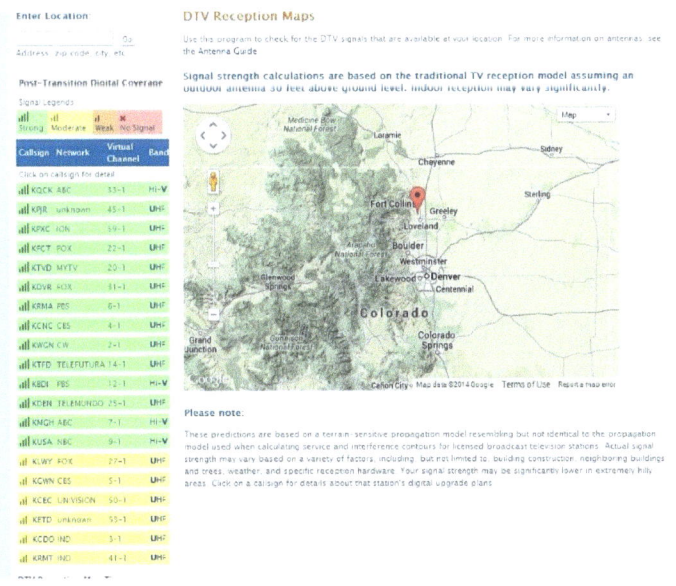

DTV Reception Maps FCC
http://transition.fcc.gov/mb/engineering/dtvmaps/

B) AntennaWeb.Org (CEA certified outdoor antennas EZ Method)
http://www.antennaweb.org/

The Consumer Electronics Association (CEA) and the National Association of Broadcasters (NAB) built this website to better select outdoor TV antennas. Their certified antennas are based on a seven color zone system. They don't really cover indoor antennas.

I am a bit weary of this information. The results that said I need a powerful rooftop antenna, I can receive with an indoor amplified antenna. The information may be old.

1. Click on *Click Here to Start.*
2. Enter your *ZIP code* in the box.
3. I recommend you put your address in the *Street Address* section. The first time I tried using just the ZIP gave me only 4 channels. Entering my ZIP code resulted in 12 channels.
4. AntennaWeb.org is mainly geared towards external roof antennas. 30 feet high and above is the ideal height for outdoor antennas. Select *"Yes"* if this is the height that your antenna will be mounted. Select *"No"* if it will be lower.
5. Click Submit, after you have entered the ZIP, Address, and antenna height.

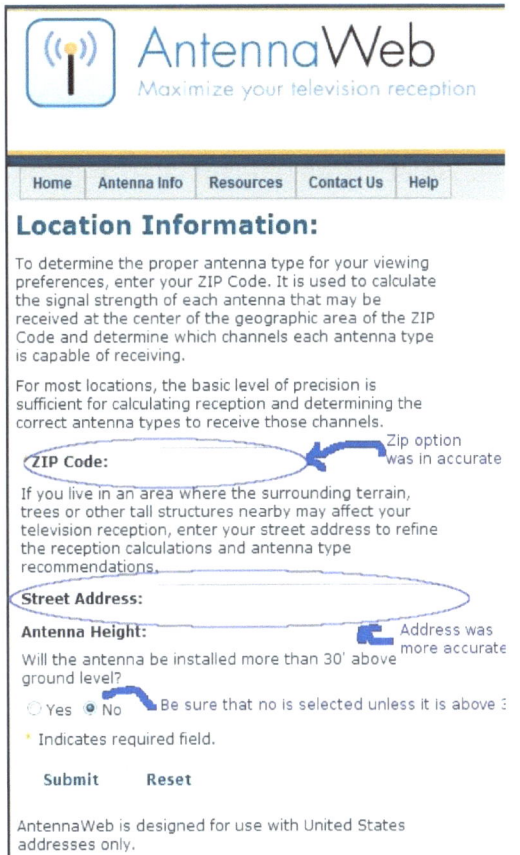

Antennaweb.org classifies the six types of outdoor antennas
*(Information from Antennaweb.org **Antenna Info** page)*

Yellow: Small multi-directional antennas – The smallest TV antenna receiving signals 360 degrees. They may be disks and patch shaped antennas or attached to satellite systems. Use in high signal areas. Yellow - 10 to 15 mile range.

Green: Medium Multi-directional – These are a bit larger and more powerful antennas that receive signals 360 degrees. These antennas may be stick, wing, or disk shaped antennas with long elements. Use this antenna when 20 foot or more runs are needed. Also if you will attach more than one device. Green - Up to 30 mile range.

Light Green: Large Multi-directional – Big size. Receives more signal power at a greater distance. Element antennas used to reject ghost situations. Light Green - Up to 30 mile range

Light Green: Small Directional - multi – This antenna is an element rooftop antenna. Light Green - Up to 30 mile range.

Yellow. Green, Light Green, Red, and Blue: Medium Directional – The most popular antenna. Multi-elemental. Amplified versions allow blue zone coverage. Red – up to 45 miles. Blue - Up 45 to 60 miles.

Green, Light Green, Red, Blue, Violet: Large Directional – Large antennas used in weak signal areas. Multi-elemental. Amplified versions allow blue and Violet color zone coverage, though will eliminate yellow zone coverage. Violet - 60 miles or more.

Where to get CEA certified antennas.

The two web stores partnering with the CEA color zone code antenna systems are:

http://www.channelmaster.com/

http://www.antennacraft.net

C) What is the distance to these transmission towers, power of these transmissions, and exact compass heading? (More Exact, Time Consuming, and Accurate Method)

Next we can figure out the distance to most of the transmission towers as well as the exact direction for compasses. We need to know the distance to help us figure out more exact what kind of TV antenna we need. TV fool.com does a great job in helping us our here. Under 25 miles, you can get away with antennas that are not amplified. Over 25 miles, amplified antennas are more desirable.

1. Go to www.tvfool.com
2. Click on **TV Signal Locator** button.

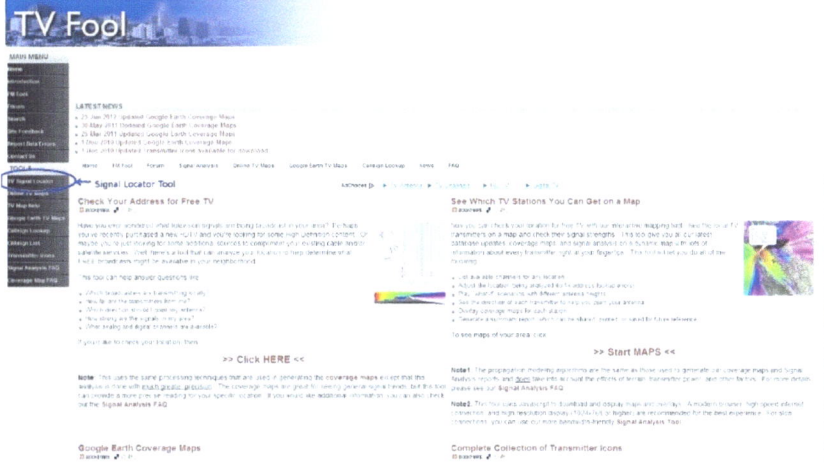

3. Enter your address and height of antenna or where you plan to put the antenna. You can search several times at different heights to see how the height will affect your signal reception.

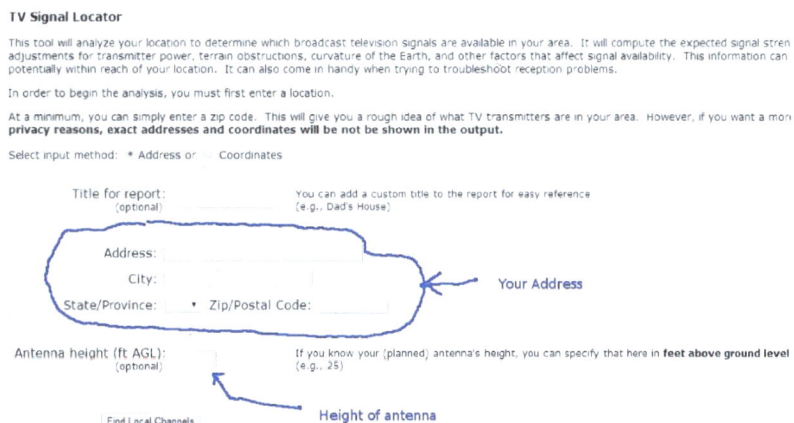

4. This method shows more information concerning signal strength, distance, and path. In addition they provide a 360 degree showing the directions of the RF channel of the towers and relative strength.

5. The All Channels channel listings on the bottom VHF Lo, VHF Hi and UHF channel lineup and strength shows the channels according to their RF channels. This VHF and UHF channel listing also shows possible conflicting channels and relative strength of each.

6. Write down the more detailed information for the channel info that you gathered from the FCC quick method concerning: real, virtual, network, distance, and direction.

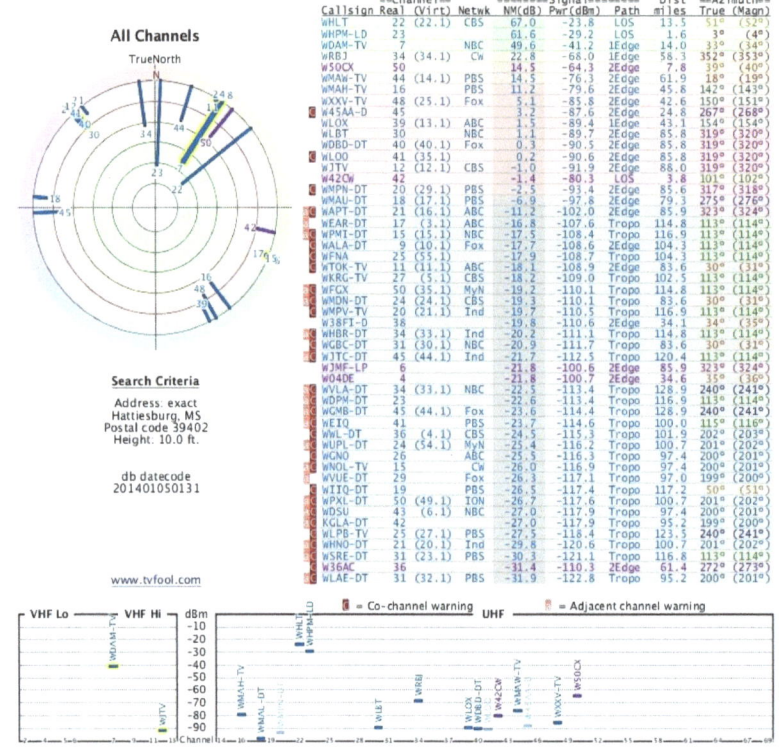

From TVFool.com

Advance: detailed list of stations and directions of towers

TVFool.com signal locator

Enter your address and height of antenna from ground level.

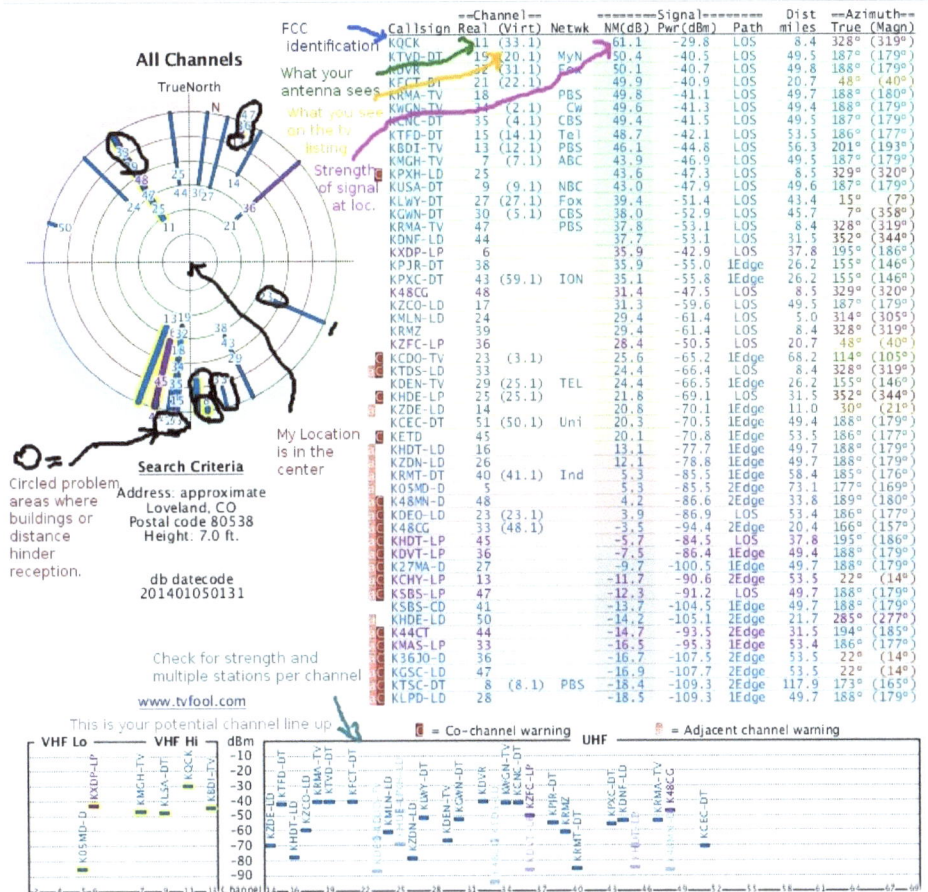

All Channels

TrueNorth

FCC identification

What your antenna sees

What you see on the tv listing

Strength of signal at loc.

My Location is in the center

Search Criteria
Address: approximate Loveland, CO
Postal code 80538
Height: 7.0 ft.

db datecode
201401050131

Circled problem areas where buildings or distance hinder reception.

Check for strength and multiple stations per channel

www.tvfool.com

This is your potential channel line up

Callsign	Real	(Virt)	Netwk	NM(dB)	Pwr(dBm)	Path	Dist miles	True	(Magn)
KQCK	11	(33.1)		61.1	-29.8	LOS	8.4	328°	(319°)
KTVD-DT	19	(20.1)	MyN	50.4	-40.5	LOS	49.5	187°	(179°)
KDVR	32	(31.1)	Fox	50.1	-40.7	LOS	49.8	188°	(179°)
KFCT-DT	21	(22.1)		49.9	-40.9	LOS	20.7	48°	(40°)
KRMA-TV	18		PBS	49.8	-41.1	LOS	49.7	188°	(180°)
KWGN-TV	34	(2.1)	Cw	49.6	-41.3	LOS	49.4	188°	(179°)
KCNC-DT	35	(4.1)	CBS	49.4	-41.5	LOS	49.7	187°	(179°)
KTFD-DT	15	(14.1)	Tel	48.7	-42.1	LOS	53.5	186°	(177°)
KBDI-TV	13	(12.1)	PBS	46.1	-44.8	LOS	56.3	201°	(193°)
KMGH-TV	7	(7.1)	ABC	43.9	-46.9	LOS	49.5	187°	(179°)
KPXH-LD	25			43.6	-47.3	LOS	8.5	329°	(320°)
KUSA-DT	9	(9.1)	NBC	43.0	-47.9	LOS	49.6	187°	(179°)
KLWY-DT	27	(27.1)	Fox	39.4	-51.4	LOS	43.4	15°	(7°)
KGWN-DT	30	(5.1)	CBS	38.0	-52.9	LOS	45.7	7°	(358°)
KRMA-TV	47		PBS	37.8	-53.1	LOS	8.4	328°	(319°)
KDNF-LD	44			37.7	-53.1	LOS	31.5	352°	(344°)
KXDP-LP	6			35.9	-42.9	LOS	37.8	195°	(186°)
KPJR-DT	38			35.9	-55.0	1Edge	26.2	155°	(146°)
KPXC-DT	43	(59.1)	ION	35.1	-55.8	1Edge	26.2	155°	(146°)
K48CG	48			31.4	-47.5	LOS	8.5	329°	(320°)
KZCO-LD	17			31.3	-59.6	LOS	49.5	187°	(179°)
KMLN-LD	24			29.4	-61.4	LOS	5.0	314°	(305°)
KRMZ	39			29.4	-61.4	LOS	8.4	328°	(319°)
KZFC-LP	36			28.4	-50.5	LOS	20.7	48°	(40°)
KCDO-TV	23	(3.1)		25.6	-65.2	1Edge	68.2	114°	(105°)
KTDS-LD	33			24.4	-66.4	LOS	8.4	328°	(319°)
KDEN-TV	29	(25.1)	TEL	24.4	-66.5	1Edge	26.2	155°	(146°)
KHDE-LP	25	(25.1)		21.8	-69.1	LOS	31.5	352°	(344°)
KZDE-LD	14			20.8	-70.1	1Edge	11.0	30°	(21°)
KCEC-DT	51	(50.1)	Uni	20.3	-70.5	1Edge	49.4	188°	(179°)
KETD	45			20.1	-70.8	1Edge	53.5	186°	(177°)
KHDT-LD	16			13.1	-77.7	1Edge	49.7	188°	(179°)
KZDN-LD	26			12.1	-78.8	1Edge	49.7	188°	(179°)
KRMT-DT	40	(41.1)	Ind	5.3	-85.5	1Edge	58.4	185°	(176°)
KOSMD-D	5			5.3	-85.5	2Edge	73.1	177°	(169°)
K48MN-D	48			4.2	-86.6	2Edge	33.8	189°	(180°)
KDEO-LD	23	(23.1)		3.9	-86.9	LOS	53.4	186°	(177°)
K48CG	33	(48.1)		-3.5	-94.4	2Edge	20.4	166°	(157°)
KHDT-LP	45			-5.7	-84.5	1Edge	37.8	195°	(186°)
KDVT-LP	36			-7.5	-86.4	1Edge	49.4	188°	(179°)
K27MA-D	27			-9.7	-100.5	1Edge	49.7	188°	(179°)
KCHY-LP	13			-11.7	-90.6	2Edge	53.5	22°	(14°)
KSBS-LP	47			-12.3	-91.2	LOS	49.7	188°	(179°)
KSBS-CD	41			-13.7	-104.5	1Edge	49.7	188°	(179°)
KHDE-LD	50			-14.2	-105.1	2Edge	21.7	285°	(277°)
K44CT	44			-14.7	-93.5	2Edge	31.5	194°	(185°)
KMAS-LP	33			-16.5	-95.3	1Edge	53.4	186°	(177°)
K36JO-D	36			-16.7	-107.7	2Edge	53.5	22°	(14°)
KGSC-LD	47			-16.9	-107.7	2Edge	53.5	22°	(14°)
KTSC-DT	8	(8.1)	PBS	-18.4	-109.3	2Edge	117.9	173°	(165°)
KLPD-LD	28			-18.5	-109.3	1Edge	49.7	188°	(179°)

🔳 = Co-channel warning   🟥 = Adjacent channel warning

VHF Lo — VHF Hi — dBm: -10, -20, -30, -40, -50, -60, -70, -80, -90

UHF

Channel

The Signal Analysis Report lists the broadcasters in your area, ranked from strongest to weakest, according to 3D of each transmitter in the table is color coded as follows:

Background color	Estimated signal strength
Green	An indoor "set-top" antenna is probably suffi...
Yellow	An attic-mounted antenna is probably needed...
Red	A roof-mounted antenna is probably needed...
Grey	These channels are very weak and will most li...

D) Sub channels (Rabbitears.info)

Another quick way to see channels and start exploring HD sub channels is by using the map of the USA at rabbitears.info. This information will also be important if you set up a TitanTV program guide.

With the information gathered from the FCC and TVFool.com, you can now figure out most of the sub channels you may also receive.

1. Go to the following links depending on chosen method.

For a quick method

Go to http://www.rabbitears.info/search.php. Enter your Zip Code at a minimum. For a more precise information, enter your address. Enter your antenna height that you plan to use, for more accurate information. These results do not take into account obstacles such as buildings or hills.

For a visual method

http://www.rabbitears.info/market.php?request=marketmap

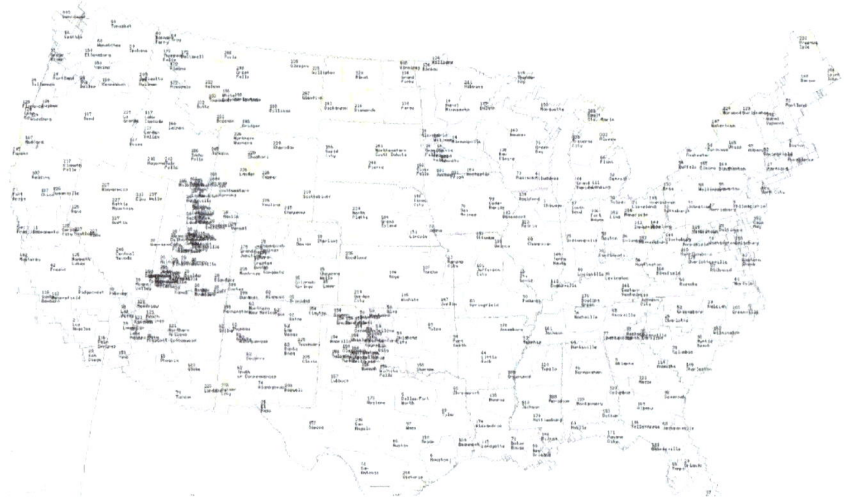

By Clicking on your market location, which you might have to locate using the pop-up label in crowded multi-viewing market areas, you will be given a list of TV channels originating and strongest within those areas. You may be able to receive stations outside this market area, but this is a good start.

2. Click on the market area you live in or are closest to.
3. *Expand All* to see a list of all the sub channels of these broadcasters.

In this example the WDAM (NBC) also transmits ABC, and Bounce TV. WMAH a PBS transmits 2 other channels: Create and MPB Music Radio. WHLT only transmits CBS. WHPM transmits FOX and CW. Digital channels can transmit up to 6 sub channels. Check with your local transmission towers to see what sub channels, besides the main channel (normally number 1 or 2) that you may pick up if you can receive and display that transmission tower's signal. The FCC page only shows the main channel.

It also gives information showing the video resolution. 480/I is SD. 1080/I is HD. Standard Definition (SD) and High Definitions (HD). Your TV must be able to receive 1080 resolution to gain the full feature effects.

Audio is either DD2.0 (2speaker) or DD5.1 (5 speaker plus subwoofer). Of course, you must have the appropriate audio equipment to hear the audio with the full feature effects.

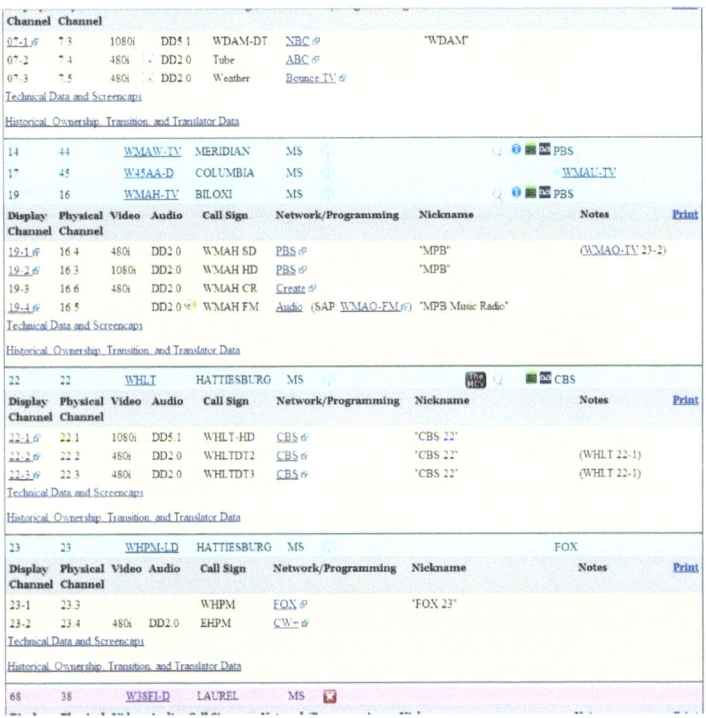

4. Go through the list of the different networks to see which networks you will receive. Here is a sample of the main networks from local TV towers in the close vicinity of Hattiesburg.

CBS [X] Chan <u>22.1-.3</u> Dir <u>NNE</u> Dist <u>14 mi</u> || Chan _____ Dir _____ Dist _____
ABC [X] Chan <u>7.2</u> Dir <u>NNE</u> Dist <u>14 mi</u> || Chan _____ Dir _____ Dist _____
NBC [X] Chan <u>7.1</u> Dir <u>NNE</u> Dist <u>14 mi</u> || Chan _____ Dir _____ Dist _____
Fox [X] Chan <u>23.3</u> Dir <u>N</u> Dist <u>2 mi</u> || Chan _____ Dir _____ Dist _____
CW [X] Chan <u>23.4</u> Dir <u>N</u> Dist <u>2 mi</u> || Chan _____ Dir _____ Dist _____
ION [ ] Chan _____ Dir _____ Dist _____ || Chan _____ Dir _____ Dist _____

Looking over this list lets me also know that I might not get ION TV station. I might check around other close viewing areas on Rabbitears to see if I might be able to pick up that channel from another close range source.

At the current time, there is also some online TV stations streamed off of sites or Filmon which can supplement and will be explained later. See those sections for further details.

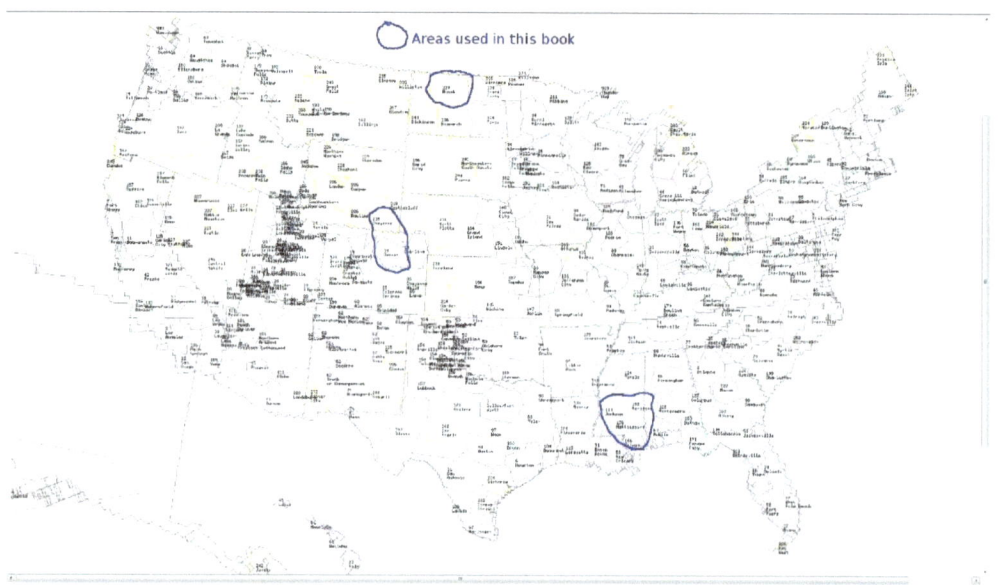

These are the areas that were checked with various tools in this book

These 3 viewing areas circled are the areas that Rabbitears covers in this book. Mississippi requires four different tower information. Colorado requires two. Antler required just one. Your viewing area may require several tower area information depending upon where you live and how many are near you.

Channel scan or auto scan

After your TV antenna is setup you need to scan for channels. Perform the following steps for most models. Read the TV instructions if these steps are insufficient.

Auto scanning for channels

1. *Make sure all cable connects are secure and input devices are turned on.*
2. *Press the Input or TV button and make sure the TV is in TV input signal mode. The mode may also have an "Ant" or Antenna mode.*
3. *You may have to continually press input button to change to antenna or TV mode.*
4. *Open the TV or converter box onscreen Menu. Select TV or Tuner option.*
5. *In the Tuner option make sure the mode is antenna, if not change it.*
6. *Select Auto search, Auto scan, Auto Channel Scan or some variation to initiate the search for channels.*
7. *When the search reaches 100% exit and use the appropriate control channel + or – button to change the channel to see what channels you have found. Use the TV remote if directly connected to the TV. Use a converter or other device if connected to that device instead.*
8. *If problems persist factory reset the TV consulting the TV your TV manufacturer's website. Normally it is done by holding down the menu or power button from 15 to 30 seconds. You can also sometime find it in the Menu under System or Setup menu.*

Scan for additional channels

Most TVs also have the ability to scan for additional channels. This ability is useful in order to add new channels whenever you change the antenna direction.

TV Program Guide

Although there is a built in electronic program guide in converter boxes and digital receivers, Using a good TV programming guide can help recording programs, checking what is on or what programs are coming up. Next I review several potential program guide to over-the-air (OTA) channels. Two program guides are pretty good though not always exact.

Each method or type of program guide I give my overall reviews of pros and cons of each type of guide.

Guide to review of program guides

- Appearance – How visually appealing the TV program guide is
- Simplicity – Easy to use, ready to go
- Convenience – How easily the guide is to access the program information
- Registration – is site registration required to use
- Grouped program categories – does the TV program guide have the ability to look for programs by program type or categories

Photo of my TV Program **Info** button results

Pressing the Info button on my built in TV remote for antenna reception brings up the channel number, channel name, program name, episode summary, TV maturity rating, time, program length by time, and TV resolution. This TV does not have a very good built in TV program guide, rather a very basic one.

Built-in-electronic program guides (EPGs)

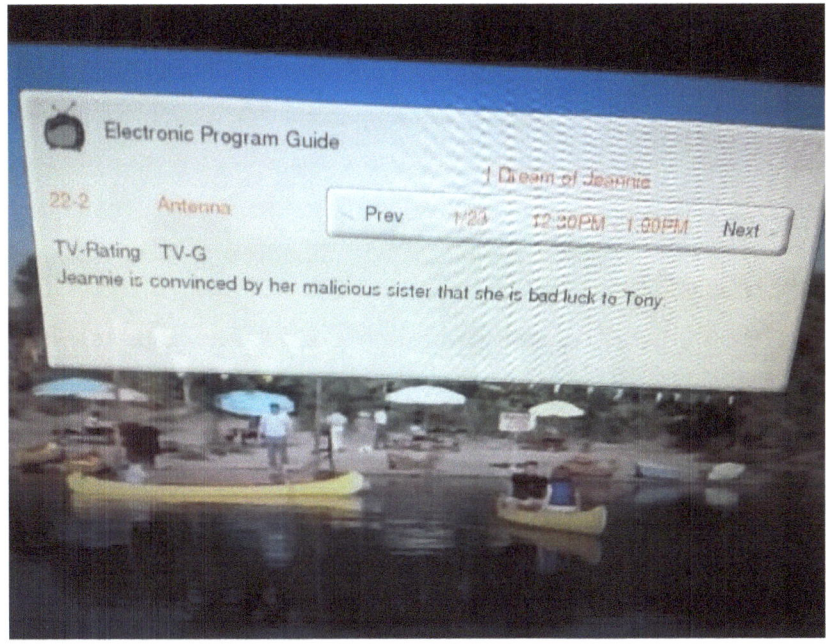

Photo of my TV **EPG** button results

Pressing the EPG button on my TV remote brings up a basic TV schedule for this antenna received channel, pressing right can show some upcoming program info. This TV is limited to

Most TVs, converter boxes, VCRs, media recorders, and computer media centers have some sort of build in program guide that shows channel shows, movies, and durations for current and upcoming content. These guides can be from a very limited time period of a few hours, to weeks in advance.

- Appearance – Varies from ok to great
- Simplicity – Most are very easy to use
- Convenience – Best Convenience. Since these are the built in to the TV, converter, software, or other receiving device, they are normally included with your main TV remote.
- Registration – No
- Grouped program categories – Not standard (some do, some don't)

Remote button normally designated to programming guide.

Accessing EPG is normally by the remote control by pressing guide, EPG, info, or some words along that kind of language. Some guides build into recordable devices VCRs and devices with hard drives have a record option.

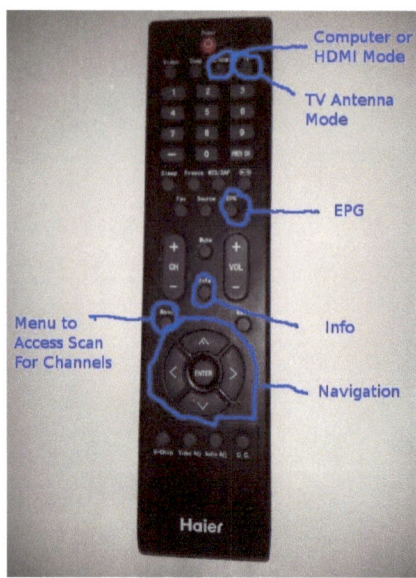

My TV remote

Navigating the guide is normally done by the arrow or direction arrows. You can normally check what is on other channels by scrolling up and down, or different times by side to side.

TVListings.AOL.com/ (Best for ready-to-go completeness)

My favorite ready to go listing is AOL's TV guide. This is powered by i.TV so it should be identical to iPhone and iPad App. If you register for a free AOL email account by just pressing the *EDIT CHANNELS* button, you can edit the channels removing unwanted listings. http://tvlistings.aol.com/

- Appearance – Great
- Simplicity – Very easy to use
- Convenience – Very Convenient
- Registration – Yes, to edit channels. No, if editing is not needed.
- Grouped program categories – Yes four groupings. Movies, News, Sports, Family.

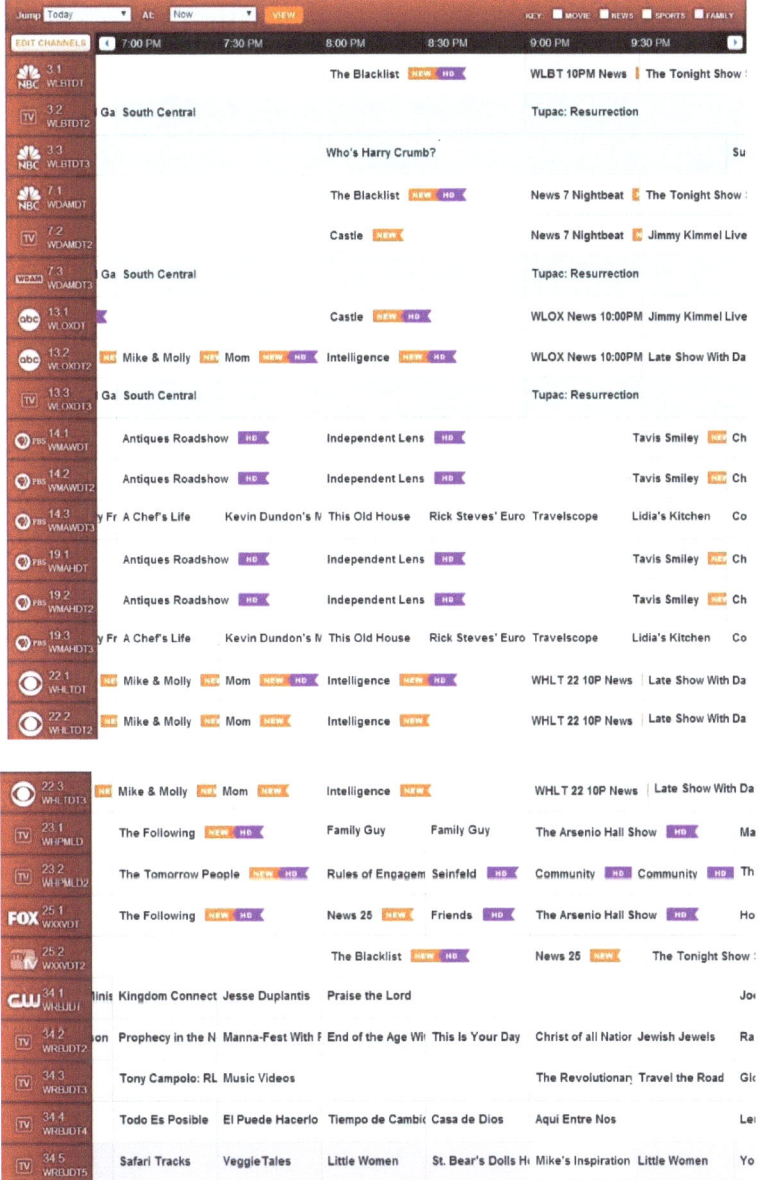

[TitanTV.com (Best for customization and details)](#)

Next TV program guide is TitanTV.com. Their guide needs a little more work to use. Once properly set up, it became one of my favorite TV program guides. I had to gather information from four different tower locations: Hattiesburg, Meridian, Gulfport/Biloxi, and a few from Jackson. It gets a little more confusing and time consuming. Each transmission area is on a separate page, so it is cumbersome. And it does not factor in confliction stations broadcast on the same channels. You can edit the Channels shown only by registering, whereas AOL is more of one stop info.

- Appearance – Very Nice when finished
- Simplicity – Complex

- Convenience – Somewhat convenient when completed.
- Registration – Yes, for customization.
- Grouped program categories – Twenty-one categories color coded. Key at the bottom.

This is a great tool after you already have an antenna and know what you can receive and what comes in. To use:

Add Broadcast Transmission Areas

1. Click the *ADD* button to create a new channel lineup.
2. Select the *Broadcast icon* to indicate that this is over-the-air transmission.
3. Enter your zip code.
4. From the drop down menu bellow your zip code, select your primary broadcast area.
5. Check the list and see if this list includes all of your receiving channels.
6. If not repeat the process, after your zip code, select another broadcast area that includes towers from a neighboring broadcast area that you receive.
7. Repeat this if necessary to include all broadcast areas your antenna receives signals from.

This process might result in many separate lists. When all are accounted for, it is time to combine data. You must register a free account to do this step.

Merge and Refine Broadcast Transmission Area Data

1. Click on the *MANAGE* button.
2. Look for the Broadcast area that has the majority of your channels that you receive.
3. Click on the *EDIT CHANNELS* button for that main area.
4. Hide channels that you do not receive or don't want program information.
5. Next click the *ADD CHANNELS* button on this same screen in the upper menu bars near the top of the channel list.
6. Choose the *Choose Channels to Add From an Existing Lineup* button.
7. Select the drop down menu down arrow.
8. Choose the next closest transmission area from the list, which you selected earlier.
9. From this list, select each of the channels that you receive which will turn green and show a (ADD) on the right. Don't add any that you do not receive nor want.
10. Click *ADD CHANNELS* button when complete.
11. Repeat process to add channels if more than 2 transmission areas, until all the channels represent what you receive.
12. You can also possibly add channels not listed in either if the information is available. I added EWTN because it was not listed in any transmission areas, although it is in our area, and I watch it occasionally. You can try ADD CHANNELs button then Find Channel to Add by Name. This may or may not work depending on available information. Or you

can try finding that channel in another transmission area, and add it to yours and see if it works.

With a little modification and configuration. Titan TV becomes the best TV program guide. Apps are available for iOS and Android devices. There is also a mobile version for other mobile device operating systems. Truly a powerful portable TV program guide.

Enjoy

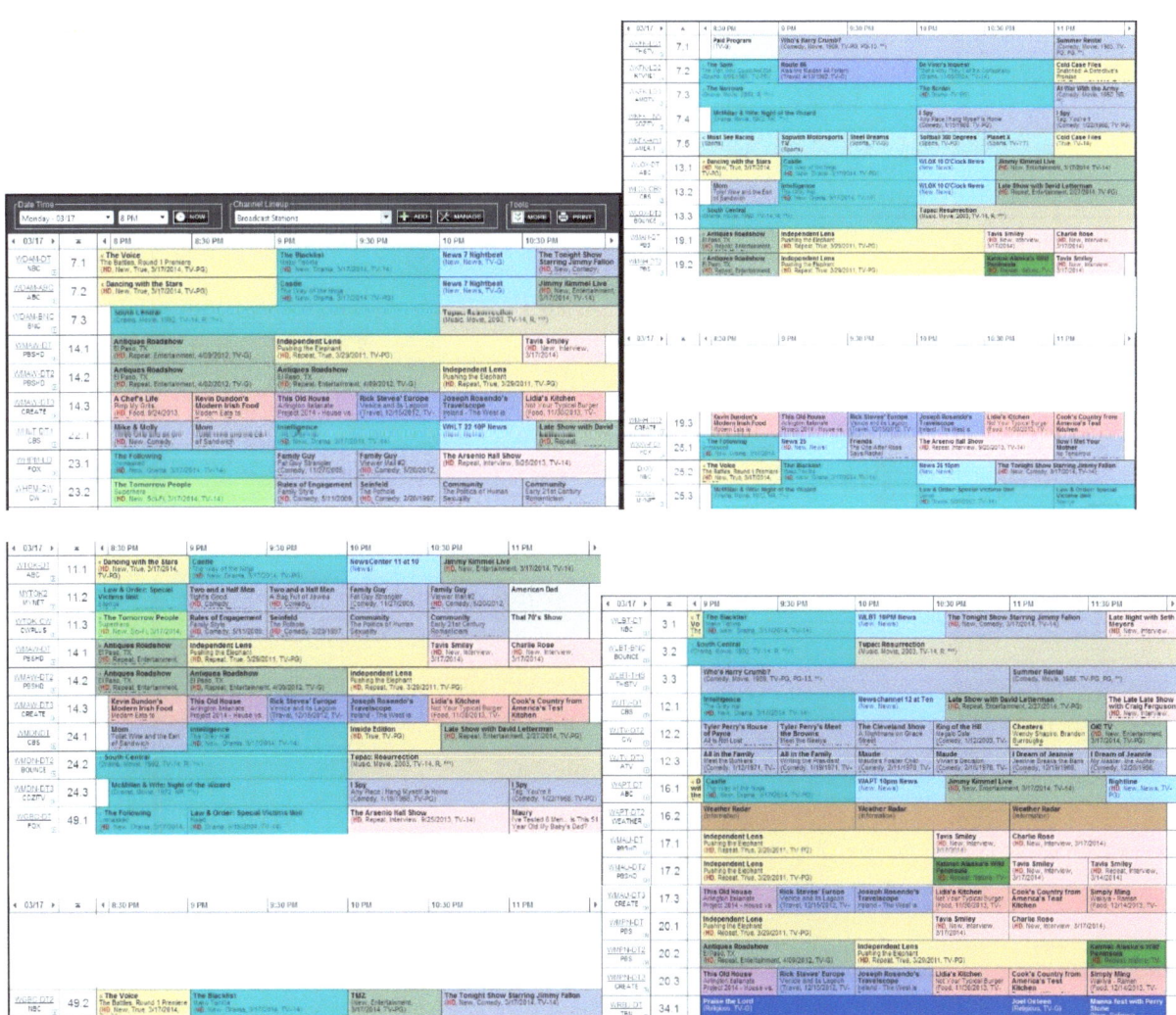

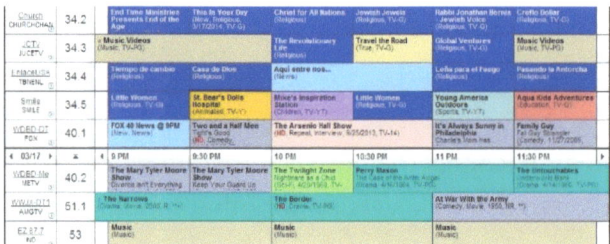

TV Guide (Most well known)

TV Guide does an ok job of bringing up the best channel, though it might miss the channels coming from further away. This guide has a tab system to keep track of movies, news, sports, family, and favorite shows. http://www.tvguide.com/

- Appearance – Plain, a bit boring.
- Simplicity – Easy to use.
- Convenience – Can be convenient.
- Registration – Yes for favorites and customization. No for as-is use.
- Grouped program categories – Five tabs; Movies, News, Family, and Sports. There also is a Favorites category.

Setting TV Guide up
1. You set it up by going to WHAT'S ON TV and then to TV LISTINGS.
2. Select the *Change The Location/Provider* link.
3. A popup will ask for your zip code. Enter your zip code and then select the *ANTENNA* button.
4. Select your Broadcast area such as *Denver Area Broadcast (Denver) (OTA Broadcast)*, which should appear after you enter your zip and select antenna.

iOS and Android devices have a TV Guide App you can install.

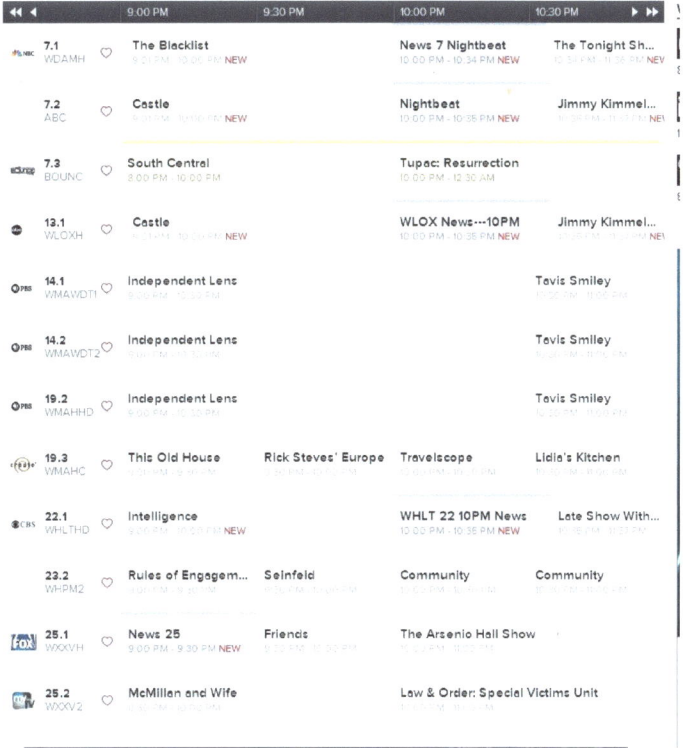

It has a tab system to keep track of movies, news, sports, family, and favorite shows.

Zap2It.com

Zap2It is similar to AOL's format, though not as colorful. I won't show all of it since it is long, filled with several ads. Yahoo uses Zap2It for its listings.

- Appearance – Very good
- Simplicity – Easy to use
- Convenience – Somewhat convenient
- Registration – Yes, for customization. No, for as-is use.
- Grouped program categories – Twelve categories called *Genres.* Just click to filter the desired categories.

Setting up your location and channel lineup

1. Change My Location
2. Enter zip code. Hit the GO button.
3. Select Local: Broadcast (Antenna)
4. You can change what channels are shown by removing ones you don't use by pressing the *Set Preference* link.
5. ADD only the channels that you watch. Repeat until you have a satisfactory list.

6. I like to check the **Additional Settings**: *Show Six Hour Grid* and *Show Only My Favorite Channels on the Grid settings.*

7. Optional: You can Display Description or Hide Description for a short summary of what each program contains.

Zap2It has a highlighter for favorites, sports, news, movies, and children shows.

- Select what types of programs you want highlighted

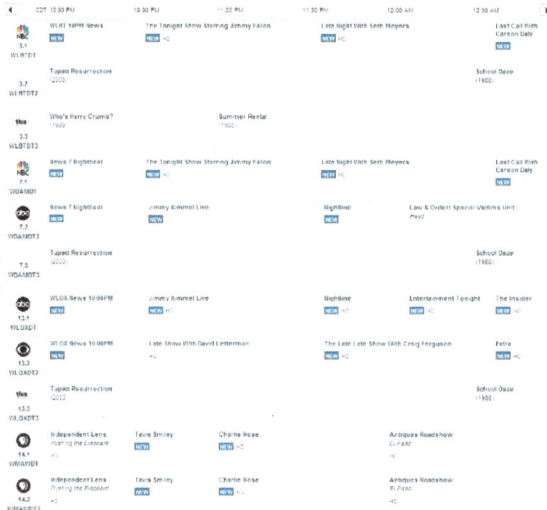

TV.com (Best for online streaming)

1. Click on LISTINGS menu heading.

2. A default area will pop up. Enter zip code. Click on the Choose Provider down arrow and choose a Broadcast TV, (Over-The-Air) appropriate.

3. You must have an account to hide channels and make favorites. To hide any channels hover over near the station number and icon. Select the "X" to any channels you want to hide. To favorite channels, do the same except select the heart symbol.

- Appearance – Similar to TV Guide.com
- Simplicity – Fairly easy to use
- Convenience – Someone convenient. Integration of show information and online viewing sources makes this service stand out among all others. This might be a favorite of online streamers.
- Registration – No
- Grouped program categories – Fourteen categories plus favorites.

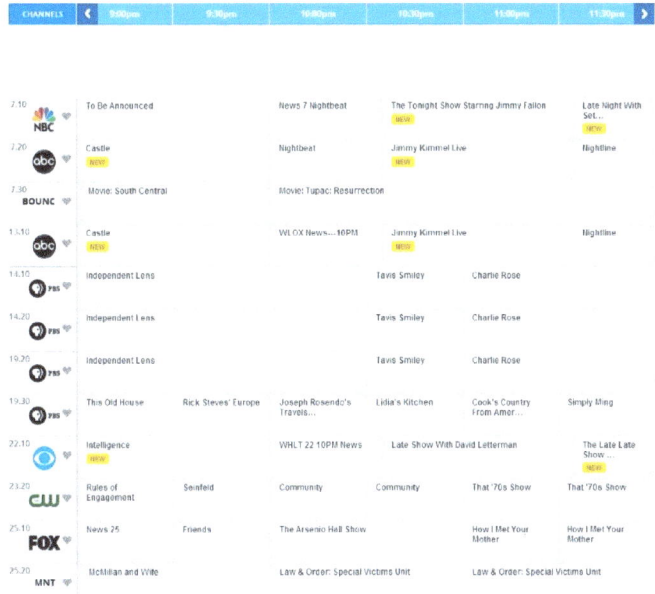

Results are a very similar to TVGuide.com except a little more colorful icons and NEW show indicator. It does not have any method to find types of TV shows, movies, news, nor sports like some of the others.

Windows Media Center

After connecting your antenna to your TV Tuner card you can open Windows Media Center, go to *TV*, and then go to *Live TV*. Normally if you have never used the service before it will start the set up to scan for channels. When this whole walk-through process is done, click on Guide to see the TV program guide.

If you have used the tuner before you may have to go to *Tasks> Settings> TV> TV Signal>*, then *Set Up TV Signals*.

Mobile Users

Check your app stores for available TV program guide apps.

Retailers that sell antennas

Store and online

Target www.target.com
Walmart www.walmart.com
Best Buy www.bestbuy.com
Radio Shack www.radioshack.com
Sears www.sears.com
Kmart www.kmart.com
Fry's www.frys.com

Home Improvement

Lowes www.lowes.com
Home Depot www.homedepot.com
Ace Hardware www.acehardware.com
Menards www.menards.com

Used

Goodwill www.goodwill.org
Salvation Army Family Stores http://satruck.org/national-family-stores
Search for local thrift shops http://www.thethriftshopper.com/

Online

Antennas Direct www.antennasdirect.com
Amazon www.amazon.com
New Egg www.newegg.com
EBay www.ebay.com
Overstock www.overstock.com
Rakuten www.rakuten.com
Think Geek www.thinkgeek.com
Crutchfield http://www.crutchfield.com/g_15920/TV-Antennas.html
Wineguard http://www.winegard.com/get-free-tv/
Clear TV (As Seen On TV 25 mile range) http://www.buycleartv.com/

Outdoor Antennas

Solid Signal http://www.solidsignal.com/
Channel Master http://www.channelmasterstore.com/
TV Antennas Sale http://www.tvantennasale.com/
Home Antenna http://www.homeantenna.org/

Basic TV Terminology

Electromagnetic waves

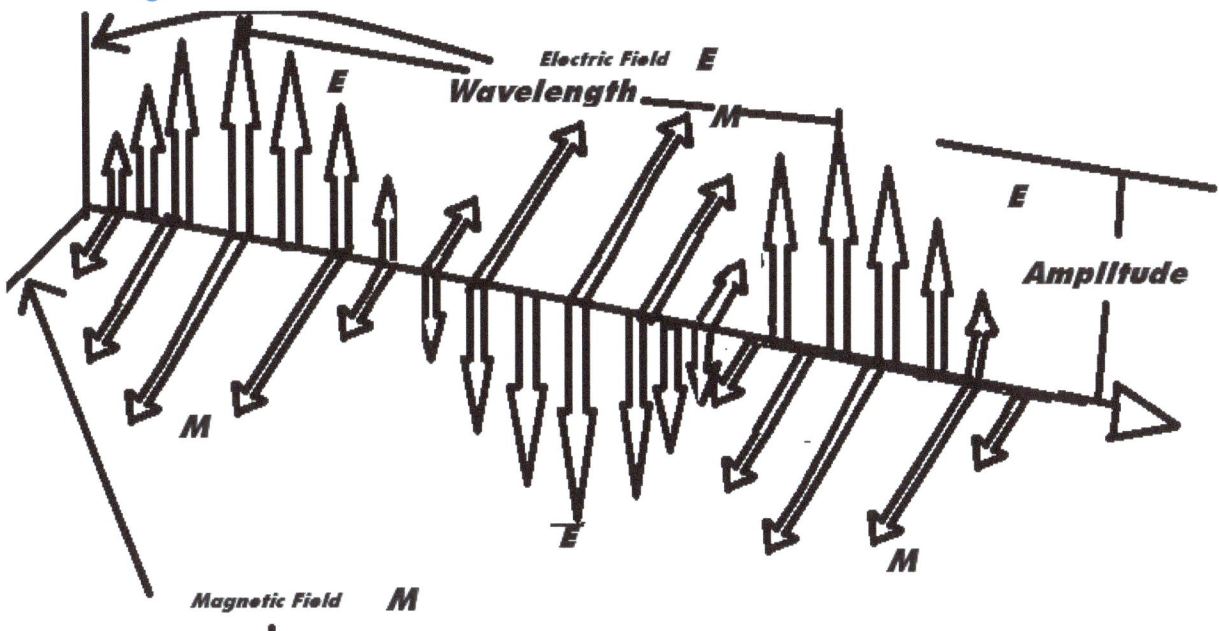

Drawing of electromagnetic wave by Ken Wickham

Antennas receive electromagnetic waves (EM) in specific spectrum. Our eyes see colors in the visual color spectrum. Radios pick up a specific spectrum of electromagnetic waves. These waves are electric fields travelling away from the source, the transmission tower. A magnetic field travels with the electric field thus making up the electromagnetic wave.

Speed of light = speed of electromagnetic waves

Both the speed of light and the speed of electromagnetic waves are roughly 671 million mph. The speed of sound is only 761mph at sea level, so the speed of waves is much faster than the speed of sound.

Sinusoidal Wave

Sinusoidal waves can be expressed by a combination of two axis of amplitude and either a) length or b) time.

Wavelength

Electromagnetic waves repeat themselves every meter (λ= Greek symbol Lambda). λ =   which means the wavelength equals the speed of light [c] divided by the frequency [v]. The speed of light per second is 186,212 miles/second.

Frequency

Electromagnetic waves repeat themselves every so often. Frequency (f = Latin F). T = time it takes to complete one cycle. Frequency is normally written in number of cycles per second called Hertz (Hz =)

Energy

Energy [E] equals Planck's constant [h] times frequency [v]. In the equation h is Planck's constant, h = 6.626 x 10-27 erg-seconds.

NOTES

Streaming Devices + Streaming Services

Reviews, comparisons, and step-by-step instructions

By Ken Wickham

Overview of Streaming section of Book

This book has been to help you know what streaming services and devices are available. The book will begin with an overall checklist of major steps to accomplish setting up services. Next, you will get to know the type of connectors, tuner, and resolution of your current TV or computer that you will be using to receive streaming services from the internet. You will figure out what content you currently own, keeping in mind the possibility of integrating that content into the streaming system. You will write down what content you desire. You will then figure out what formats, providers, and devices that content may be available, through using mainly three free online internet search services. My device recommendations for different situations will be presented. Then brief information will be presented for major streaming technology and devices available currently. With this information, you may begin to figure out your own recommendation if you would rather figure it out on your own, rather than following my suggestions. A few feature summary tables will be presented, as well as summary of device prices, contents, hardware, and setup steps for each of the devices. I give my overall steps to selecting streaming services in an attempt to replicate pay TV to an extent. I then give my streaming service recommendations for various situations. Streaming services will then be introduced with a brief summary of available content, prices, and popular content. After that, sports packages will be presented with the associated costs and brief major features. Finally, lists of major broadcast TV and Pay TV online internet sites with links/addresses will be presented with a small summary of available content such as streaming live content, streaming episodes, and clips.

The book intends to help guide you in setting up your own internet streaming services. It does not cover older model streaming devices, however there may be some similarity with manufacturer newer models. The goal is to replace your pay TV massive bill with smaller fees, more customized content, and near free yet legal.

Step to receiving streaming service

1. [   ] First write down the qualities of your TV or computer monitor you will be watching. This information includes connector types, TV tuner, and screen resolution. Also record your internet service speed. Record this info on the planning information sheet in section I. You may print off a copy of the planning pages.

2. [   ] If desired, fill out the movie inventory sheet(s) listing all of your current movies. I provide one that you can use by printing off as many needed copies. On that sheet, I make available room to record other information that may help you keep track of your movie library. This may help you in deciding how to integrate existing content into your streaming services and overall entertainment system.

3. [   ] Decide what type of streaming content you wish to see. You can do section 2 filling out your favorite content, and using the three free online search website tools to fill out most of the content supplier availability information. See the instructions, for quicker help gathering where these TV shows and movies are located.

4. [   ] If you will base your content on a device you already own, what is the content limited to that device? Use the charts at the beginning of the streaming device section to compare content available on popular devices. If you will be purchasing a device, make sure to consider the limited content available to the devices during your decision process. Make sure you can receive and use the content which you wish to use. If you have no priority of content, realize some content you might not be able to receive on different devices.

5. [   ] If needed, purchase your device either based on price, desired content, features, speed, TV connections, and content reception options.

6. [   ] Set the device up to the internet service.

7. [   ] Set up the content you already possess.

8. [   ] Purchase, subscribe, or configure the streaming services that you will be using. Subscriptions and services may offer some free time or number for downloading.

With this overview of the steps needed to plan, analyze, and set up your streaming services, we will first look more at a series of pages you can print off, with this checklist, to help you do steps 1 through 3.

For a printable PDF of these steps and charts, I make them available off of my Google Drive. Click on the link or direct your browser to this address.

https://drive.google.com/file/d/0B2I_GEXbFycdOE4zWG9SLUYyMm8/view?usp=sharing

Planning your TV entertainment services

I. Streaming Device and Streaming Service

[Print these pages off if needed] See the chapter "What Kind of Viewing TV or Computer Tuner do you own?"

In the book or on separate pieces of paper, please write down the following information in order to help you setup an alternative to expensive paid TV

What type of television connectors does your TV have? How many?

- Coaxial _____
- HDMI _____
- S-Video _____
- Composite (LR(red white) audio, (yellow) video _____
- Component (Green - Y, blue – Pb, red – Pr; all three are video, audio is separate)
- USB _____
- VGA _____
- DVI _____

What kind of TV tuner will you be using? (Check or write down each one)
- D. Built in TV   [ ]
- E. TV converter box   [ ]
- F. Computer TV tuner   [ ]

What is your TV resolution?
- F. 480i (analog) [ ]
- G. 480p (digital) [ ]
- H. 720p [ ]
- I. 1080p [ ]
- J. WQHD 1440p [ ](Highest currently possible)

What is your Internet download speed? Search Google "*ISP name* speed test", filling in your ISP provider.

_____Mb/s

Streaming Internet Speed minimum recommendations

(Circle your internet speed range)

Laptop 1 Mb/s
SD TV 1-2 Mb/s
HD TV 720p 2-4 Mb/s
HD TV 1080p 5-9 Mb/s

II. What TV shows, series, and movies do you watch?
www.tv.com, http://www.canistream.it/ , www.eTRIZZLE.com

1.) Fill out the first form, the list of content on the sheet on the following page, which you currently watch or plan on watching. Start with sports packages if desired, then movie/TV shows.

Which internet sports packages do you want or already have?

NBA	MLB	NHL	MLS	NFL	UFC	MMA	WWE	NASCAR	WNBA

Next you can make a list of all the content that you already own. I have created an inventory list that you can use to record this and other important information. Some of these providers only provide live audio some provide live internet. Check the sports section to see which are live, and which are next day.

- If you have a numbering system, you can number your movie or show content in the first column

Movie – You can put the name of the movie or show in the 2nd column. If you are organized you can use any form of organization such as alphabetical, by genre, by acter/actress, etc...

Part in Series – I created this column for movies or shows in some sort of series. It can be a movie series or TV show season/episodes. You might want to indicate in this column something such as "Season 6" or episodes 1 to 4, or part 2 in a trilogy.

Genre – Here you can put categories such as Sci-Fi, Comedy, Romance, Action, Adventure, Fantasy, Anime, Horror, or Documentary to give you a few examples.

Minutes – Length of the movie, episode, show, DVD, or whatever.

Rating – This can either be something like PG, R, PG-13 or better yet your own rating system such as 1 to 5 star or A through F. It can be something simple like good, great, sucks, hated.

Format – Is this ownership content in the form of which of these categories

Location – Where is this content located? Insurance may want a list if your movie collection is destroyed in some disaster or fire/flood/tornado.

Value/Cost/Price – Insurance may want a list of the value of your movie/show collection if perishable.

2.) Next, fill out the 2nd form of content desired, listing shows, series, and movies that you wish to see. Use the three *Content Search Tool Website.* See that section for details and instructions.

www.tv.com, http://www.canistream.it/ , www.eTRIZZLE.com

#	Movie	Part in serie	Genre	Minutes	Rating	Format	Main Actors/Actresses	Value/Cost/Price	Location
						Blu-ray◯ DVD◯ Digital◯ VHS◯			
						Blu-ray◯ DVD◯ Digital◯ VHS◯			
						Blu-ray◯ DVD◯ Digital◯ VHS◯			
						Blu-ray◯ DVD◯ Digital◯ VHS◯			
						Blu-ray◯ DVD◯ Digital◯ VHS◯			
						Blu-ray◯ DVD◯ Digital◯ VHS◯			
						Blu-ray◯ DVD◯ Digital◯ VHS◯			
						Blu-ray◯ DVD◯ Digital◯ VHS◯			
						Blu-ray◯ DVD◯ Digital◯ VHS◯			
						Blu-ray◯ DVD◯ Digital◯ VHS◯			
						Blu-ray◯ DVD◯ Digital◯ VHS◯			
						Blu-ray◯ DVD◯ Digital◯ VHS◯			
						Blu-ray◯ DVD◯ Digital◯ VHS◯			
						Blu-ray◯ DVD◯ Digital◯ VHS◯			
						Blu-ray◯ DVD◯ Digital◯ VHS◯			
						Blu-ray◯ DVD◯ Digital◯ VHS◯			
						Blu-ray◯ DVD◯ Digital◯ VHS◯			
						Blu-ray◯ DVD◯ Digital◯ VHS◯			
						Blu-ray◯ DVD◯ Digital◯ VHS◯			
						Blu-ray◯ DVD◯ Digital◯ VHS◯			
						Blu-ray◯ DVD◯ Digital◯ VHS◯			
						Blu-ray◯ DVD◯ Digital◯ VHS◯			
						Blu-ray◯ DVD◯ Digital◯ VHS◯			
						Blu-ray◯ DVD◯ Digital◯ VHS◯			
						Blu-ray◯ DVD◯ Digital◯ VHS◯			
						Blu-ray◯ DVD◯ Digital◯ VHS◯			
						Blu-ray◯ DVD◯ Digital◯ VHS◯			
						Blu-ray◯ DVD◯ Digital◯ VHS◯			
						Blu-ray◯ DVD◯ Digital◯ VHS◯			
						Blu-ray◯ DVD◯ Digital◯ VHS◯			

TV Show or Movie	Website	Netflix	Hulu	Hulu Plus	Amazon Prime	Amazon Instant	HBO GO	VUDU	iTunes	GooglePlay	Crackle	YouTube	Popcorn Flix	MGO	CinamaNow	PLEX	EPIX

What Kind of Viewing TV or Computer Tuner do you own?

The first step you need to do is to figure out if your TV has a built in digital TV tuner, or is an older analog TV tuner. If you are going to use a computer as a TV, you will need to check if you have a TV tuner card (normally indicated by a coaxial input port on your computer).

Built into TV [ ]
Converter Box [ ]
Tuner Card [ ]

At the same time, you may want to make a list of all the devices you plan on including in your entertainment system.

Accessories
Antenna [ ]
Media Player [ ]
Media Streamer [ ]
DVR [ ]
DVD player [ ]
Blue-ray player [ ]
Video game console [ ]
Sound system or speakers [ ]

TV Connectors

Next, you need to know what type of connectors are included on your television(s) you will be using.

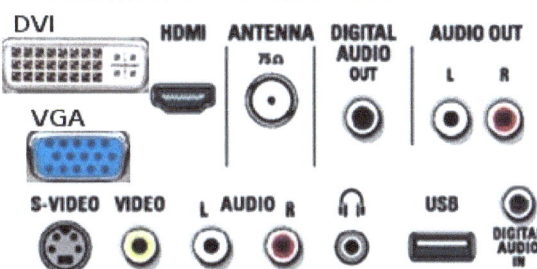

Common TV Connectors

HDMI

HDMI emerging in 2004, HDMI has replaced coaxial as the high definition connection of modern televisions. 90% of HDTVs by 2007 had HDMI connectors. By 2009, all digital televisions had at least one HDMI connectors. Many converters exist to convert different type of connectors and wires into HDMI. The signal probably will not play at the highest resolution however. Be cautious of having too many connectors, splitters, and wires which leads to signal loss and overall masses of wire octopuses.

S-Video

S video has a max 480i/576i signal definition. S-Video is slightly better than composite video, using 2 channel encryption instead of one. You may have to use this for older TVs though composite a/v is normally more common.

Composite A/V

Older TVs might need to use the composite port. Composite has a max of 480i/576i. This is the 3 color wires of yellow for video, and red and white for audio. Composite A/V is slightly lower quality compared to S-Video because it only uses one channel instead of the two that S-Video uses.

Ethernet

Cat-5e is the current standard for Ethernet wiring. Ethernet wiring is mainly used to connect devices by wire to routers and modems.

Coaxial

Older still is the basic coaxial input and output. RG-6 is the current standard for coaxial cable.

USB

USB port on TVs can run movies, listen to music, and look at pictures from a thumb drives. It can also update the TVs firmware by downloading the software putting it on a thumb drive then inserting the drive into the port. External hard drives can also be used to play content connecting directly to the TV.

TV resolution, aspect ratio, formats

When the entire United States switched over from analog to digital service, the entire country was thrown into a little frenzy. What emerged however has its advantages and disadvantages.

Modern signal types

Broadcaster digital terrestrial television (DTV) broadcasting. Broadcasters normally transmit one or both types of picture formats, which vary in size and aspect-ratio. High definition television (HDTV) for the transmission of high-definition video and standard-definition television (SDTV).

The i and p in the resolution

In the signal transmission the gap is called *interlacing video* (symbol = i) in the analog method and *progressive* (symbol = p) for digital video. In interlacing they sent half a frame at a time, first the odd lines then the even lines. Now they send line-by-line. So when you read 720p, this actually means 1280x720 digital *progressive scan* signal in that it receives a signal line by line. This reduced eyestrain from interline twitter making images and movement more smooth.

Old analog signals (for comparison)

CGA computer monitors had a resolution of 320x200 up to 640x200 4 bit 16 color.

VGA computer monitors (1987) introduced 640x480 up to 800x480 16 bit in 4:3 aspect ratio. VGA has 256 colors in 320x200 mode. The second resolution is the VHS and Beta-max resolution. 480i is also the broadcast resolution of old analog televisions.

Current HDTV signals

HD 480p has 640x480 pixel or 4:3 aspect ratio (4 units wide by 3 units high) is SDTV. This is closest to DVD quality which is 720x480 which can be shrunk to fit SDTV, or play in wide-screen mode with dark space above and below the picture.

HD 720p widens the resolution to 1280x720 16:9 aspect ratio.

HD 1080p resolution of 1920 × 1080 at a 60 Hz with 16:9 aspect ratio. This is Blue-ray resolution.

With a few pages of content plans, a little knowledge about TV connections, resolution, and wire connections, we can now look into the basics of streaming. We can then look at some of the popular streaming devices. We can compare device features, characteristics, advantages, and disadvantages, and prices.

What is streaming video? And what are the main basis of streaming services?

Content Search Tool Website

www.tv.com, http://www.canistream.it/ , www.eTRIZZLE.com Tools to search multiple online content providers.

COMPARISON OF CONTENT SEARCH TOOLS

Search	Hulu Plus	Netflix	Amazon Instant	Amazon Prime	Redbox Instant	Redbox Kiosks	YouTube	Crackle		Popcornflix	VUDU
tv.com	√	√	√	√							√
CanIStream.It	√	√	√	√	√	√	√	√			√
eTRIZZLE.com	√	√	√	√	√	√		√			√

Search	MGO	iTunes	CinamaNow	PLEX	EPIX	Google Play	Xbox	Sony Entertainment	Target Ticket	Snag Films
tv.com		√				√				
CanIStream.It					√		√	√		√
eTRIZZLE.com	√	√				√	√	√	√	

This table is shown to show you which search website will search which provider services. Notice that none of these three searches CinamaNow, PLEX, or Popcornflix. Use these to fill in the second sheet of desired content. I show an example for searching for a TV show, in this case HBO's Game of Thrones, and a movie, in this case Resident Evil.

1.) TV.com

1A.) TV SHOWS/EPISODES

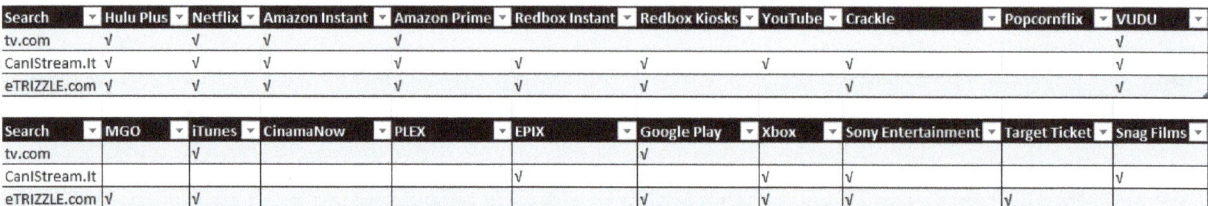

To help you fill out the rest, go to www.tv.com, 1) click in the *Search TV.com* box and type the title. Find the show or movie logo or picture, then look for and click the **Watch** link near the bottom of that section.

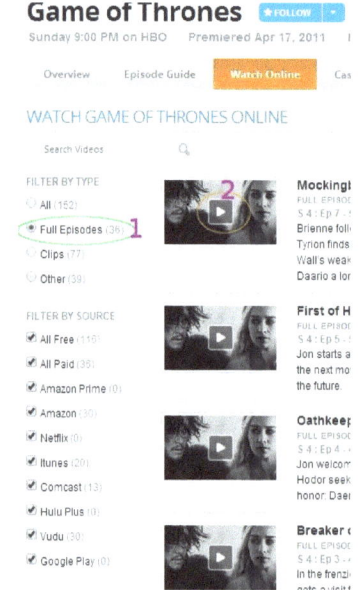

1.) Click on the Full Episodes filter for TV shows. 2) Hover over play button for play options (hover shows results below). Options are FREE, SUBSCRIPTION, or BUY IT for TV show episodes.

1B.) MOVIES

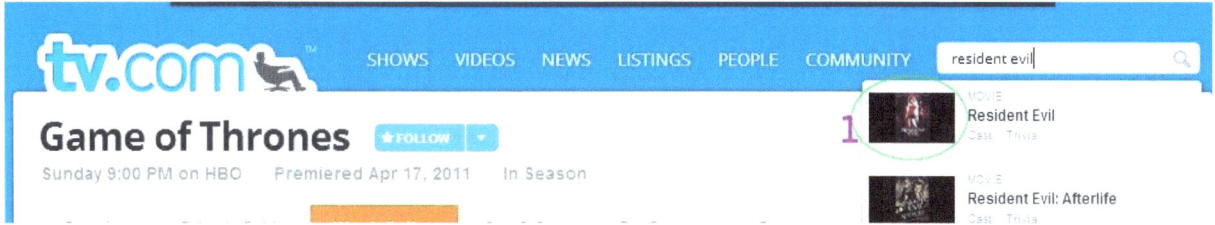

For movies, 1.) you can click on the movie in the result section.

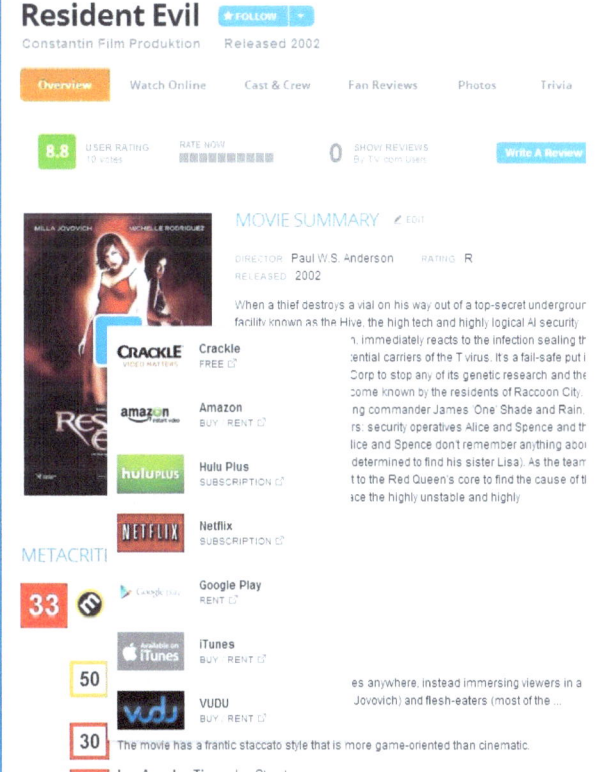

2.) Then hover over the **play** button. Although this does not give ALL the available options, it is a start. It gives the content for Amazon Prime, Amazon Instant, Netflix, iTunes, Hulu Plus, VUDU, and Google Play which are the big guns of streaming services. It will note if the method is FREE, BUY, RENT, or SUBSCRIPTION based from each of the providers.

2.) CanIStream.It

Next, you can also use http://www.canistream.it/ which searches four main areas usable by cord cutters: streaming, digital renting, digital purchase, and disc purchase/rentals.

For **streaming,** it searches Netflix, Amazon, Hulu Plus, as well as Crackle, YouTube, EPIX, Xfinity Streampix, and Snag films.

For **digital renting,** it searches Amazon, iTunes, Google Play, VUDU, YouTube, and Sony Entertainment Network, Target Ticket.

For **digital purchasing**, it searches Amazon, iTunes, Google Play, VUDU, Xbox 360, Sony Entertainment Network, Target Ticket.

For **disc Purchase and Rental**, it searches Amazon (both DVD and Blue-ray), Netflix, and Redbox.

Also Xfinity subscribers can check packaged channels through CanIStream.It

2A.) TV Shows Episodes

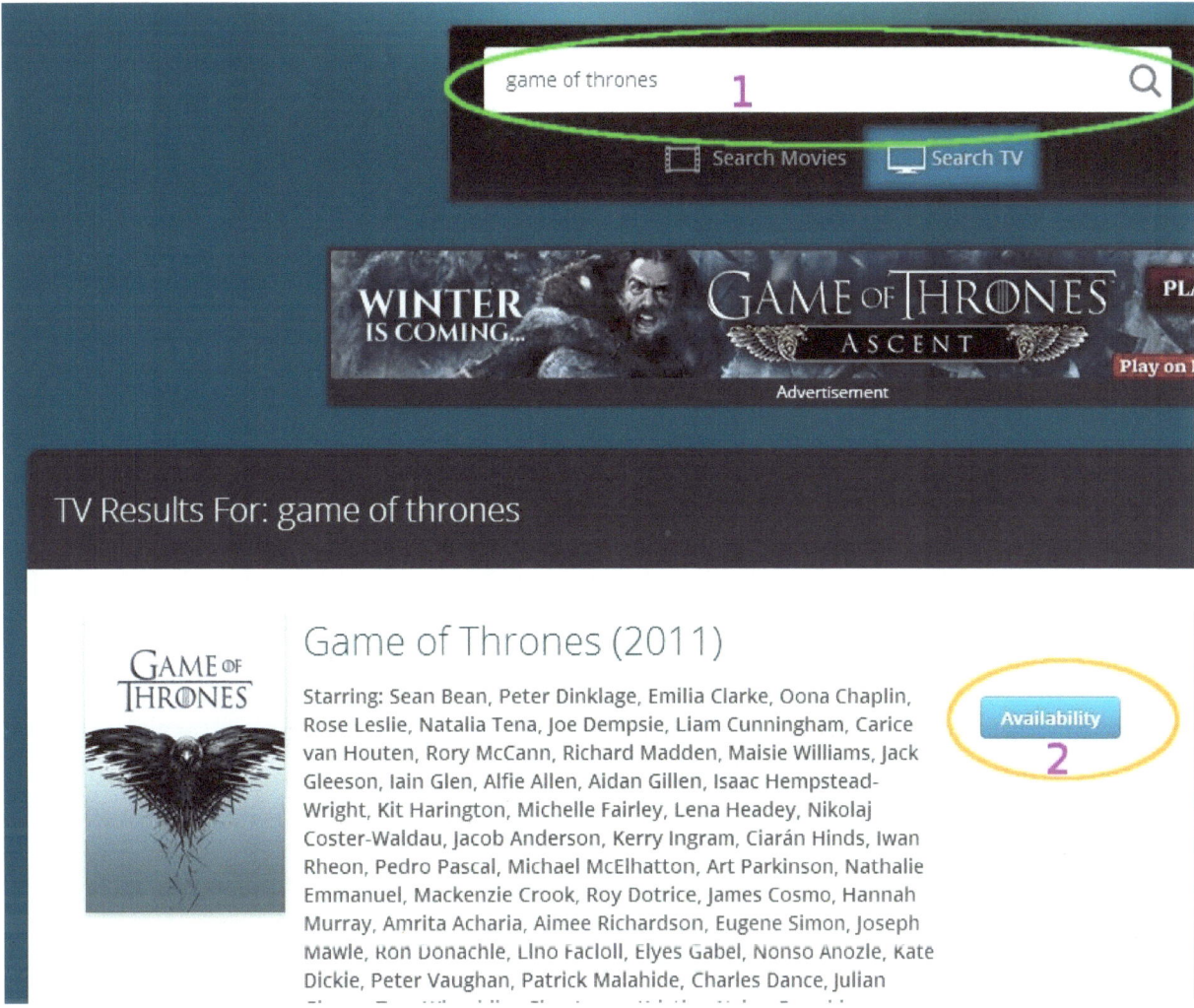

1.) First enter the title in the search box. Next, be sure to indicate underneath whether it is a **Movie** or a **TV** program. Click on TV for this search.

2.) Click on **Availability**.

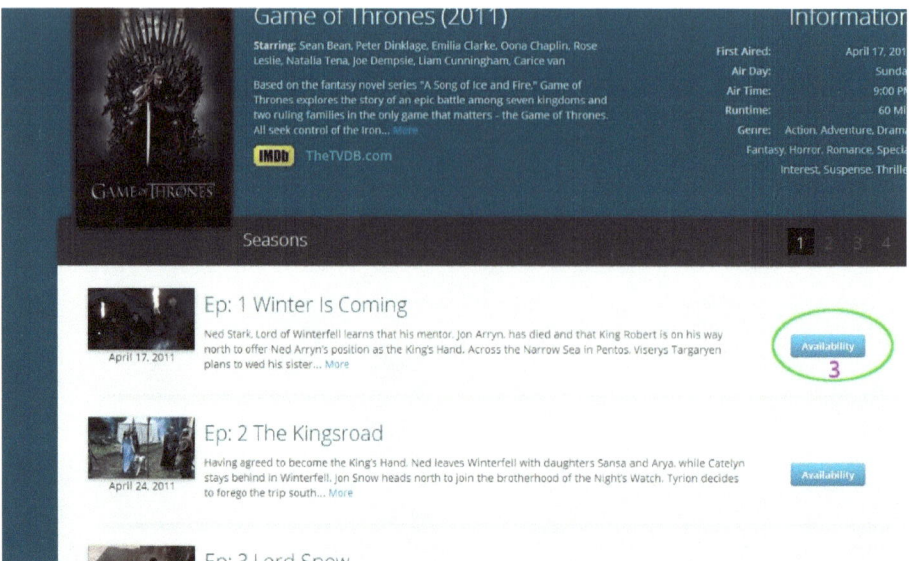

3.) Look for the episode that you are wanting to watch. Some episodes will not be available on some formats until the end of the season, especially for pay TV shows. Click Availability on the episode that you want to see.

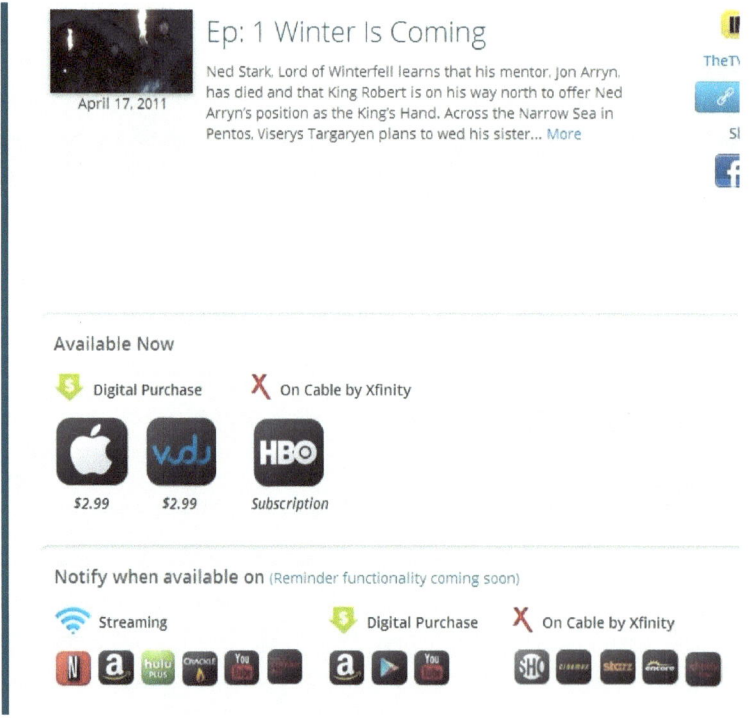

4.) The icon will show Streaming prices, ads for free streaming, or subscription and digital purchase prices. At this moment you cannot rent episodes. Also you can select Notify When available on select services, if signed in.

3.) Click on the service icon to go to that service.

2B.) MOVIES

For movies the search is quicker.

1.) Type the title of the movie in the search box. Make sure Search Movies is selected below the search box.

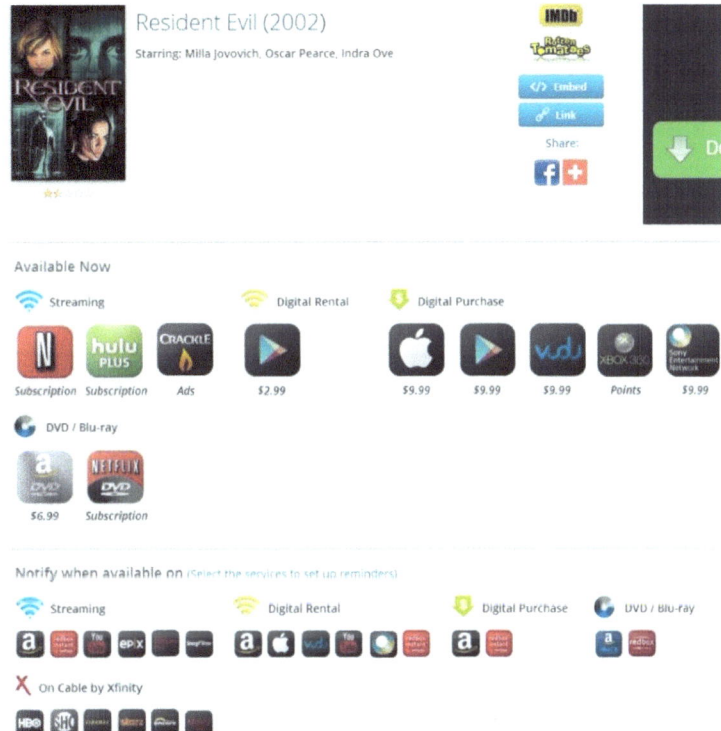

2.) It will return the results listing streaming options. Digital rental, digital purchase, and DVD/Blu-Ray. Also you can select Notify When available on select services, if signed in.

Prices will be shown for purchasing/rental. Free will mention Ads, Xbox uses points. Subscription services will be mentioned below the symbol.

3.) Click on the service icon to go to that service

3.) eTRIZZLE.com

eTrizzle searches TV shows from iTunes, Amazon Instant, Hulu, Netflix, Amazon Prime, Crackle, Google Play, and VUDU.

Movies will search **subscriptions** from Netflix, Amazon Prime, Hulu, and Crackle.

Movies will search **rentals** from iTunes, Amazon Instant, Google Play, VUDU, Target Ticket, Xbox video, Sony Movies, MGO, and Redbox.

3A.) TV Shows and Episodes

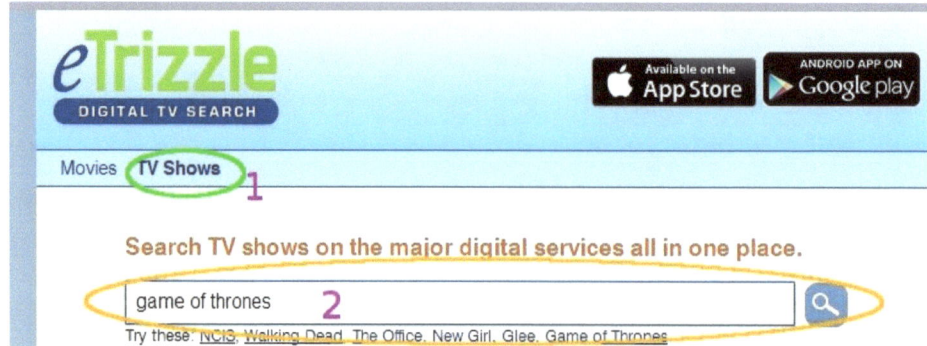

1.) Select TV Shows.

2.) Type in the search window the name of the TV show or episodes you are looking for.

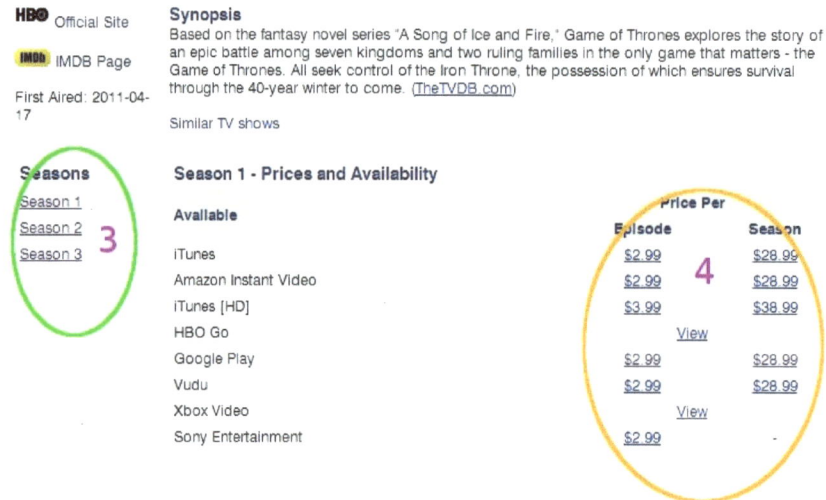

3.) Although the latest season is listed on top, scroll down to see past seasons. Click on which season you want.

4.) Available subscription, free, and purchase of episodes will be indicated. Click on the desired service indicate to the right of the service provider.

3B.) MOVIES

1.) Make sure **Movies** is selected.

2.) Enter the title of what movie you will look for in the search box. Press enter or click the search icon.

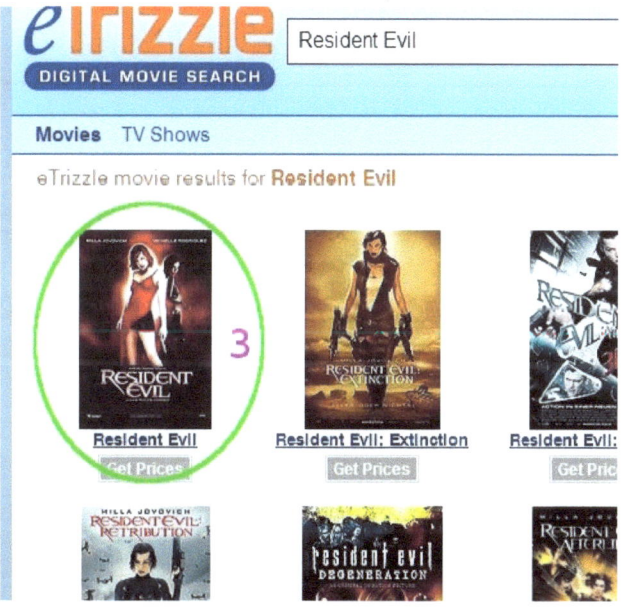

3.) If multiple titles have the same name, click on which movie out of the list is the one that you are trying to find.

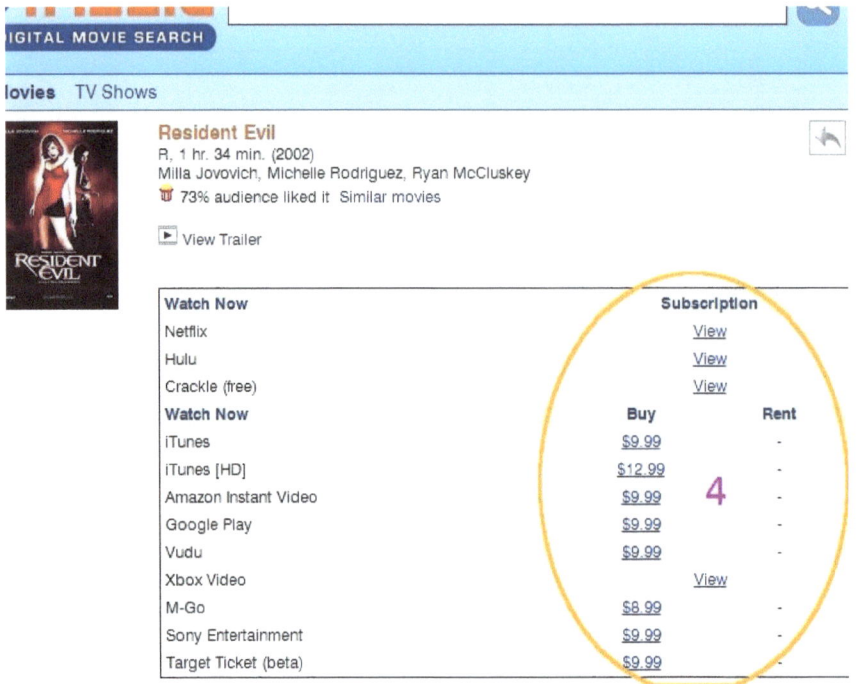

4.) Look at the available services. Select which one you prefer.

4.) Watchily

http://watchi.ly/

This will search for content according to some of what is available for your device. You just enter the name of the show or movie, select your device. Then click **Go**.

It searches Amazon Instant, Netflix, HBOGO, iTunes, MAXGO, Showtime Anytime, XFINITY

Devices supported: Roku, Amazon Fire TV, Apple TV, Chromecast, Google TV, iPhone, Android, Smart TVs, Video game consoles - PS, Xbox, Wii

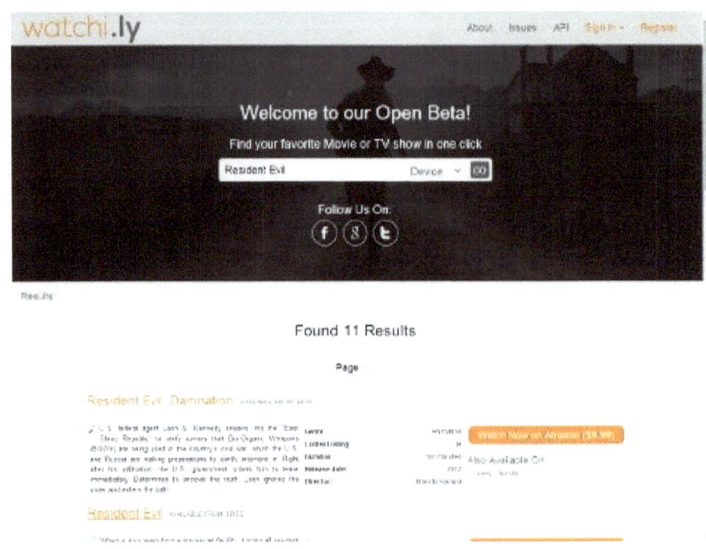

Streaming Video

Streaming video is something anyone can do with a high speed internet connection. There are three main basis of streaming services: commercial, subscription, and on pay-as-you-go.

Income base of streaming services

Commercial Based

Some networks offer commercial based on demand episodes and movies. These commercial based services ingrain multiple commercials per viewing.

Subscription Based

Paid subscription based services charge a fee per month. The paid subscription services offer a catalog of available movies which change over time.

On Demand, Pay-per-viewing, Pay-as-you-go

Instant video allows you to rent and buy video episodes and movies on demand.

With a basic knowledge of the three basis for streaming services, we can focus more on the devices. I will next give my recommendations for a few major types of devices for different situations.

My recommendations for Streaming Device Choices

Cheapest route: Use preexisting used devices. A Laptop or old computer can be kept near the TV and used for entertainment. A HDMI wire connection is preferable, although component, VGA, and S-Video connections are also doable. The only caution is sometimes older computers can't play streaming video smoothly, so a trial is needed prior to setting things up. Go to Hulu.com and try some of the streaming content first.

Next, if you have a SmartTV then your content is already build in. You can activate and use those services.

If you have a Wi-Fi Blue-ray player, then you are already to get service. Activate and use those services.

If you have one of these video game consoles you can use it without additional cost, except the Xbox and other subscription costs.

Best overall device: Roku 3 or a Laptop hooked up to your tv. The amount of content is more massive than any other device except a computer.

Fastest current device: Amazon Fire TV. The quick voice search and responsiveness of the device makes this a great device if speed is important to you. Check the device content for its limits to make sure it covers all of your desire content.

Best device for older TVs: Roku 2. Especially if you have a TV that can use composite connections. It has the most content and Wi-Fi remote at a 600 MHz CPU speed.

Best device for iTunes content users: Apple3 TV. If you have already bought content for you iPhone, iPad, or iPod, then this is the device that you want.

Best device for modifying: G-Box Q Kodi, G-Box Midnight MX2 XBMC or Android Mini PC device.

Best device to hook up to hard drives: First, WD TV Play, WD TV Live or WD TV Live Plus. Second, VIZIO Co-Star or VIZIO Co-Star LT (although for cord cutters the integration pay TV features will not work on the VIZIO).

Best device USB dongles:

Content - Roku stick, *Speed*- Amazon Fire stick, *Customization*- Android Mini-PC stick, *Price*- Chromecast or Amazon Fire stick

Streaming Devices

Figure out if you already have a streaming device that you can use to connect your television to internet content. Using a preexisting device is the most economical way of trying out streaming, without adding extra cost. You can purchase a device if getting one is absolutely necessary, or a luxury. However, when trying to save money, try to use a device you already own first.

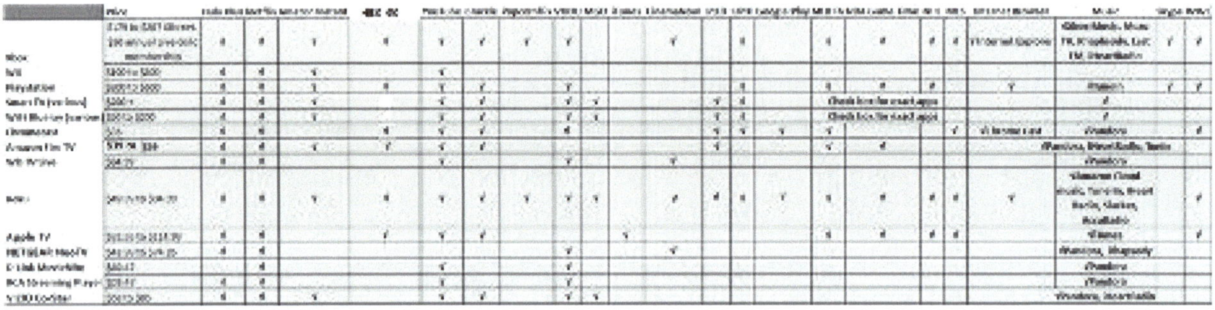

Streaming Devices you may already own

Advantages: Low cost, you may already own
Disadvantages: Bulk, not pleasing to the eye
Cost: free up to HDMI cable $20 (or other device to TV connector)

Computer, tablet, smartphone, laptop

Hook up existing computer, tablet, smartphone, or laptop to Television preferable via HDMI. It will mostly depend on what connectors the television and the computer have built in.

Different HDMI connectors. HDMI regular size, HDMI Mini (Tablets probably use most), HDMI Micro (Smartphones).

HDMI Wire

The biggest drawback is the inconvenience of having to hook up and unhook the computer if it is not dedicated to the television. Also the computers can be big, bulky, and awkward, in addition to not fitting in with the overall appearance of most entertainment systems.

The tablets and smartphones I have hooked up may get quite warm and use up power quicker if plugged into a charger.

Change TV mode

Normally when you connect your computer, tablet, console, or streaming device to your computer you will need to change the mode or source for.

	Price	Hulu Plus	Netflix	Amazon Instant	HBO GO	YouTube	Crackle	PopcornFlix	VUDU	MGO	iTunes	CinemaNow	PLEX	EPIX	Google Play	MLB TV	NBA Game Time	NHL	MLS	Internet Browser	Music	Skype	WWE
Xbox	$179 to $287 XBoxes, $50 annual Live Gold membership	✓	✓	✓	✓	✓	✓	✓	✓					✓	✓	✓	✓	✓	✓ Internet Explorer	✓Xbox Music, Muzu TV, Rhapsody, Last FM, iHeartRadio	✓	✓
Wii	$100 to $400	✓	✓	✓		✓	✓	✓	✓													
PlayStation	$200 to $600	✓	✓	✓	✓	✓	✓	✓	✓	✓	✓			✓	✓	✓	✓			✓Tuneln		✓
Smart TV (various)	$200+	✓	✓	✓	✓	✓	✓	✓	✓	✓	✓	Check box for exact apps		✓	✓					✓	
Wii Blueray (various)	$50 to $200	✓	✓	✓	✓	✓	✓	✓	✓	✓	✓	Check box for exact apps		✓					✓	✓	
Chromecast	$35	✓	✓	✓	✓	✓						✓					✓	✓Chrome Cast	✓Pandora	✓
Amazon Fire TV	$39 to $99	✓	✓	✓	✓	✓	✓	✓		✓	✓							✓Pandora, iHeartRadio, Tuneln	
WD TV Live	$84.99	✓	✓		✓	✓	✓	✓	✓		✓							✓Pandora	
Roku	$49.99 to $94.99	✓	✓	✓	✓	✓	✓	✓	✓	✓	✓	✓	✓	✓	✓	✓		✓Amazon Cloud music, Tune In, iHeart Radio, Stacker, AccuRadio	✓
Apple TV	$92.95 to $114.99	✓	✓	✓	✓	✓				✓				✓	✓	✓	✓	✓	✓iTunes	✓
NETGEAR NeoTV	$43.99 to $74.85	✓	✓	✓	✓	✓				✓				✓	✓	✓			✓Pandora, Rhapsody	
D-Link MovieNite	$40.47	✓	✓	✓	✓	✓												✓Pandora	
RCA Streaming Player	$59.47	✓	✓	✓	✓	✓												✓Pandora	
VIZIO Co-Star	$58 to $86	✓	✓	✓	✓	✓	✓	✓										✓Pandora, iHeartRadio	

Video Game Consoles

Advantages: You probably already own it

Disadvantages: Some consoles like Wii have very limited content.

Cost: No more cost if you already own it, just cost of subscriptions. Xbox currently has gold membership annual requirement cost.

Many People already enjoy gaming on video game consoles. Don't buy these units just for their streaming ability. They are priced extremely too high if bought only for their streaming ability. Buy them for their video game content first. Secondary is the ability to play Blue-rays and DVDs. Third is the ability to stream. The three main consoles have the ability to stream through apps.

Xbox

Prices and pictures

360 $163 to $300

One $350 to $600

Content

Current streaming apps include Hulu Plus, Netflix, Amazon Instant, HBO GO, ESPN, Encore Play, FX Now, Fox Now, History, Lifetime, Showtime Anytime, Starz Play, VUDU, Target Ticket, EPIX, Red Bull TV, NBA Game Time, Popcornflix, Crackle, SnagFilms, iHeart Radio, MTV, Dailymotion, and YouTube. To see the apps that you can use to stream see the official 360 or One site. Most apps require additional subscriptions fees in order to access content. http://www.xbox.com/en-US/LIVE/partners

Speed

Xbox 360 has a 3.2 GHz triple core 512 MB RAM, with a 500 GHz GPU. Xbox One has a 1.75 GHz eight core 8 GB RAM, with an 853 MHz GPU. The 360 is said to have problems overheating sometimes. The One is built with cooling in mind with a 10 year lifespan constantly powered on.

Setting it up

1. Xbox with the Xbox live gold membership ($50 currently) and Wi-Fi internet you can stream several streaming services to your television. You may have to pay additional for paid streaming services such as Hulu Plus monthly. Purchase your desired streaming service first from their website. Write down your username/email and password(s)

2. Xbox Live Gold Membership http://www.amazon.com/Xbox-LIVE-Month-Gold-Membership-One/dp/B0029LJIFG

3. Sign in with your Xbox Live Gold Membership enabled gamertag.

4. Go to the App section.

5. Select *Browse Apps*. Select the desired video streaming app(s) to download them. Limit is your memory space.

6. To start the app go to *Video* or *TV & Movies*, then go to **My Video Apps**.

7. Select the one you wish to start.

Wii

Price and picture

Wii $129
Wii U $250

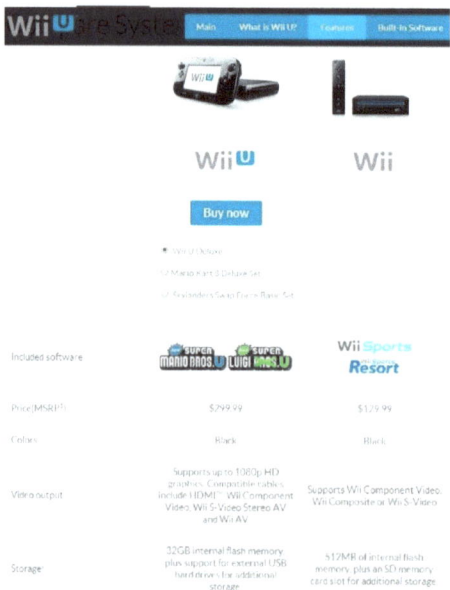

Content

Amazon Instant Video, Hulu Plus, Netflix, YouTube

Speed

Wii has a 729 MHz processor and 512 MB RAM, with a 243 MHz GPU. Wii U has a 1.24 GHz triple core and 2 GB RAM, with 1GB of the 2GB RAM shared between the CPU and the GPU.

Setting up internet connection

1. Power up. Press the "A" button on the remote to access the main menu. Select the "Wii" button with the Wii remote.
2. Select "Wii settings".
3. Select the right arrow ">" to move to page two.
4. Select "internet".
5. Then select "connection settings".
6. Choose wired or wireless instructions based on your situation.

A) Wireless

1. Click on "wireless connection"
2. Make sure there is a clear spot, clear one, and select "Connection 1: None".
3. Next press "Search for an Access Point".
4. Press "OK".
5. Search for your wireless connection.
6. Select your wireless connection.
7. Enter password or key, then select "OK". If it connects you are connected. If not, troubleshooting is needed.
8. Click "Save Settings" and then "Yes"

B) Wired

1. Install Wii LAN adapter connected to the USB port.
2. Connect one end of the Ethernet cable into the LAN Adaptor and the other into the Modem or Router.
3. Click on "wired connection".
4. Select "OK".
5. Wii will attempt to connect to the internet. If successful, press "OK". If not, some troubleshooting is needed to figure out why the system will not connect.

Make sure the system is updated

1. From the main menu of the "Wii Settings"
2. Select the right arrow ">" twice to move to page three.
3. Select "Wii System Update."
4. Select "Yes".
5. Review system update message. Press "I accept" if you agree with terms of the message.

Download the app

1. Select the Wii Shop icon from the Wii Menu.
2. Select "Start Shopping".
3. Select the desired app.
4. Select "yes, keep running," if present with a memory space message.
5. Select "No" if you are not a current subscriber, select "Yes" you already are.

Playstation – PS3, PS4, PS VITA

Prices and pictures

$199 to $270 PS3

$399 to $500 PS4

$200 PSVITA

Content

Current streaming apps include Hulu Plus, Amazon Instant, Netflix, HBOGO, VUDU, EPIX, MLB TV, NBA Game Time, NHL, WWE Network, TuneIn Radio, Crackle, and YouTube.

Speed

PS3 is reported to have a 3.2 GHz CPU and 256 MB RAM, with a 550 MHz GPU. PS4 has a 1.6 GHz 8 core shared with APU and 8 GB RAM, with an 800 MHz APU. PSVITA is reported having a 2 GHz 4 core processor and 512 MB RAM, with a 200 MHz 4 core GPU.

Internet connection [Easy mode]

1. Setting up internet connection
2. First make sure you have your access device name and password/key before starting for your wireless connection.
3. Power on your modem and router (if you have one). Wait until all lights required lights are on especially the internet and Wi-Fi.
4. In the XBB main menu go to Settings (icon looks like a tool box).
5. Go to *Network settings* and press the **X** button.
6. Go to *Internet Connection* and select [Enabled].
7. Scroll down to *Internet Connection Settings* and press the **X** button.
8. Select the [Yes] button if it says it will be disconnected.
9. Select [Easy] and press the **X** button

*Follow the wireless or wired procedure.

A) Wireless

Pre start up: Make sure no Ethernet is connected to the PS.

1. At the *Connection Type* screen, select [Wireless] and press the **X** button.
2. At the *WLAN Settings* screen, select [Scan] and press the **X** button.
3. From the list, find your SSID (for modem or router), select it and press the **X** button.
4. Confirm your SSID then press your right button. Don't press **X** button at this moment, which will edit your SSID.
5. Select your Security Type, which is normally WPA or WPA2.
6. Enter your security key or password for the modem or router.
7. Press the right direction arrow.
8. Once all the info is entered, a list of settings will be shown. Press the **X** button to save these settings.
9. Press the X button to test the connection.
10. The test will succeed or fail. A success means that the wireless connection is ready.

B) Wired

Pre start up: Before starting make sure the Ethernet is connected to the PS and the router or modem.

1. After pressing the X button from step 9 of the Easy connection mode, the system will check for your configuration.
2. A list of settings will appear.
3. Press the **X** button to *Save* the settings.
4. Next *Test* the connection by pressing the **X** button.
5. The connection will succeed or fail. A success means you are connected.

Using a Streaming Service

1. If the service requires a subscription, first go subscribe to the service using their website. Take note of any username and passwords.
2. Some free services require at least registration.
3. Connect your PS to the internet.
4. Sign on to your PlayStation Network account. You may have to register if you have not done so previously picking your username and password.
5. Go to your **XMB** (*Xross Media Bar*).
6. Go to your *PlayStation Network* section on the bar.

Installing Streaming App

7. Go to the XMB and find the TV/Video section.
8. Download the desired app.
9. Sign in to the app with your username/email and password.
10. Enjoy

Smart TV

Possible Brands: Samsung, VIZIO, Sony, LG, Panasonic, SHARP, TOSHIBA

Price range
$170 to $3000+

> Advantages: All-in-one, compact
> Disadvantages: Higher cost, limited apps

Content
Most Smart TVs have: *Amazon Instant Video, *Hulu Plus, *Netflix, **VUDU, Crackle, **M-Go, YouTube, *requires paid subscription. **requires pay for each movie or episode

Setting up
Wi-Fi connection: Just need to hook it up to your network. Several services require subscriptions indicated below.

Blu-ray Player
Wi-Fi enabled Blu-ray players with apps.

Samsung, VIZIO, Sony, LG, Panasonic, SHARP, TOSHIBA

Price range
$50 to $250

Content

Like the Smart TV, most Wi-Fi Blue-ray Players have:

> ** Amazon Instant Video, *Amazon Prime, *Hulu Plus, *Netflix, **VUDU, Crackle, **M-Go, YouTube,

*requires paid subscription

**requires pay for each movie or episode

Buy a Streaming Device

Chromecast

http://www.google.com/chrome/devices/chromecast/

Price and picture

Chromecast by Google ($31.59) + Chrome Browser (Free)

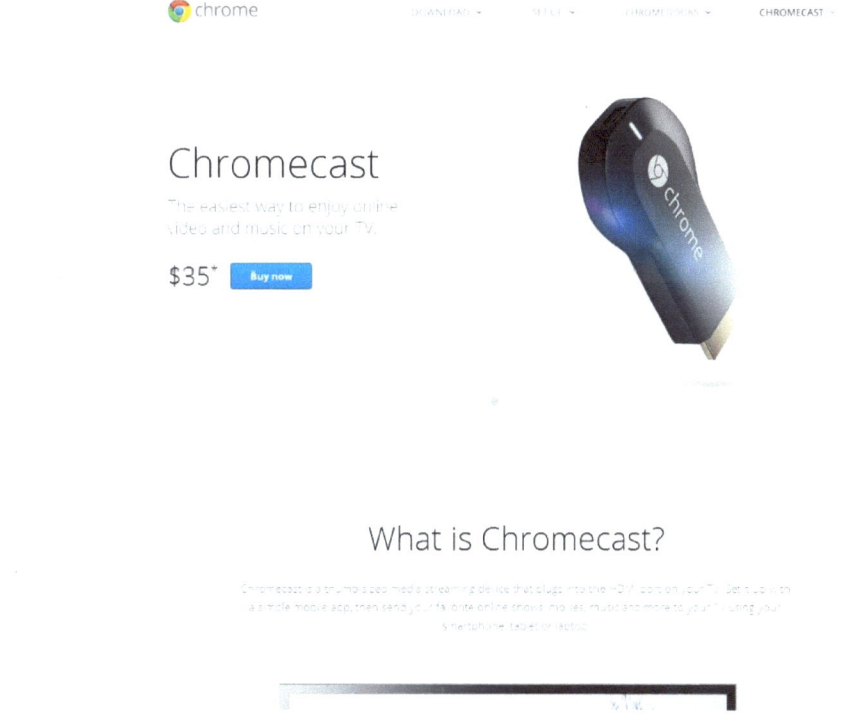

Streams many apps that you cast from your phone, tablet, or Chrome browser on computer. Casting means you send your current internet position to the Chromecast device connected to your TV which it retrieves.

> Advantages: Low cost, small, some Chrome casting ability. Pretty improved content.
> Disadvantages: It lacks some major content providers such as Amazon, iTune, and TuneIn, has pretty weak games,

Content

*Netflix, YouTube, *Hulu Plus, Crackle, Pandora, iHeartRadio, Rdio, Google Play Movies and Music, PlutoTV, *HBOGo, *Starz Play, *Showtime Anytime, EPIX, VUDU, SnagFilms, TED, Filmon, Dailymotion, *MLBTV, MLS Matchday, NFL GamePass, *WatchESPN, *UFCTV. RedBullTV *requires paid subscription

Speed

Chromecast is said to have a 1.2 GHz CPU and 512 MB RAM + 2GB Flash, and Vivante GC1000 GPU.

Setting it up

1. Install the correct version of Google Chrome browser for your operating system. https://www.google.com/intl/en/chrome/browser/

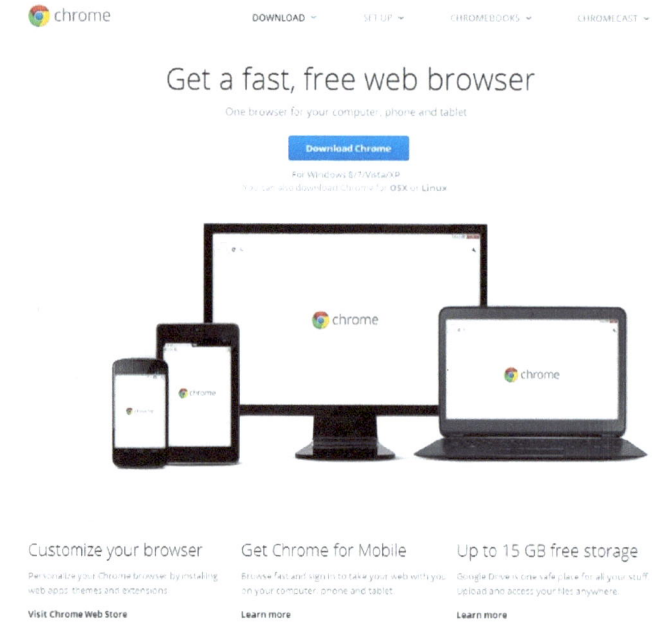

2. Open your chrome browser.
3. Search for "chrome extensions" in the browser or go directly to the extension page at https://chrome.google.com/webstore/category/extensions .
4. You search for the "Google Cast" in the web store search area.

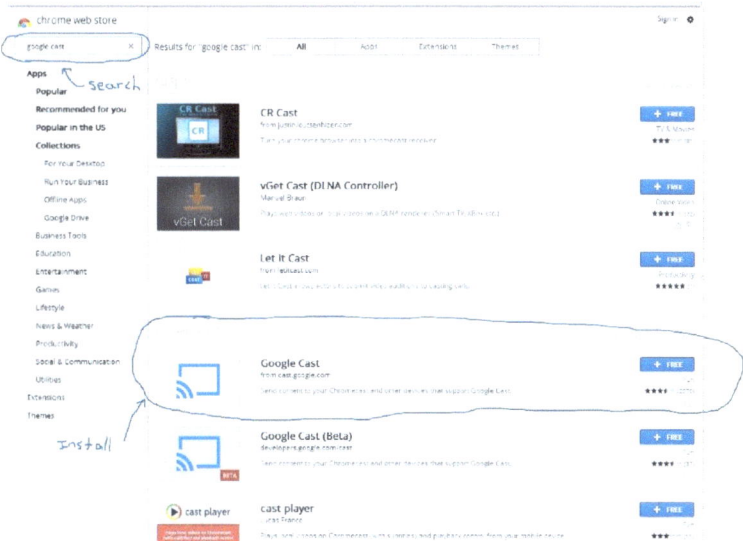

5. Install the free Google Cast extension from cast.google.com.

Hooking up Chromecast

1. Plug the Chromecast into a HDMI port in your television.
2. Plug the other end into either a USB port in your television or into the included power charger leading to a wall socket or power grid.
3. Power on the television.
4. Select the appropriate HDMI as the source or mode of your television.
5. With your chrome browser that you installed, go to the website http://google.com/chromecast/setup with your laptop, phone, or computer when it tells you to.
6. It will also tell you what your Setup Name is: _____
7. You will **download** the Chromecast application. **Install** it.
8. Start the application you downloaded. It will search for your Chromecast device.
9. When it finds your device, or a list of Chromecast devices select yours that you will be using for this TV and **Continue**.
10. Letters and numbers should now appear on both your TV and computer/laptop/phone.
11. Click on **That's My Code** to continue.
12. It will then ask for your network name and password for your wireless network.
13. Then you name your Chromecast device
14. Click **Continue**.

To Cast from the Chrome browser

1. Open the browser. Press the cast button in the browser.
2. Click it again if you want to pause supported videos or music, or disconnect.

Amazon Fire TV & Fire Stick
www.amazon.com/FireTV

Price and picture

$99

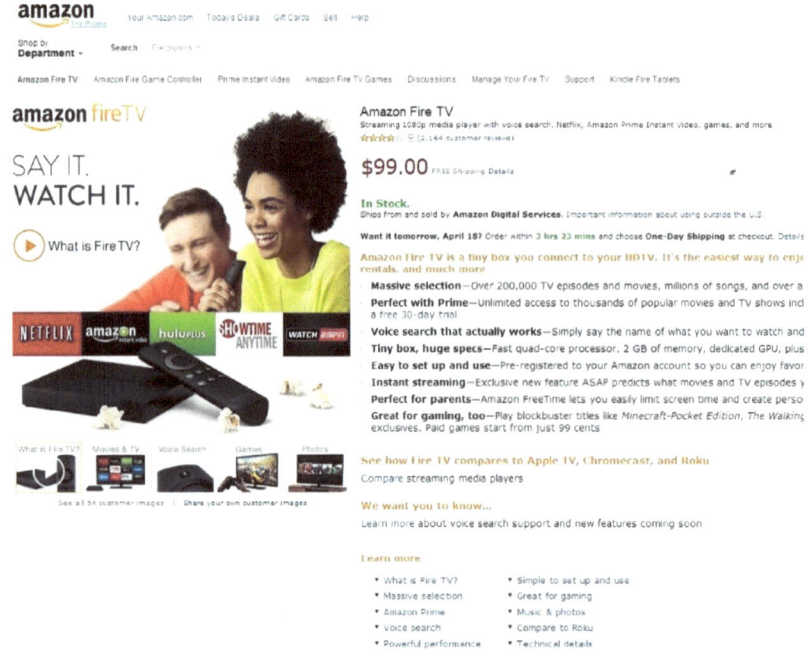

Fire TV Stick - $39

New from amazon is an integrated streaming device for TVs.

> Advantages: High speed, voice search,
> Disadvantages: Higher cost, It lacks a few major content providers such as Google Play, VUDU,
> iTune, and most sports packages

Contents

Netflix, Amazon Instant Video, Amazon Prime Instant Video, Hulu Plus, Flixster, Watch ESPN, Showtime Anytime, YouTube, Crackle, RedBullTV, PLEX, Pandora, HBOGo, TuneIn, Spotify, iHeartRadio, Amazon Music, PlutoTV, TubiTV, WWE, Weather Live, Weather4Us

Speed

Amazon Fire TV has a 1.7 GHz quad core CPU and 2 GB RAM, with a 400 MHz GPU. It has 8 GB Storage

Amazon Fire TV Stick has a 1.0 Ghz dual core CPU and 2 GB RAM, with a VideoCore GPU. It has 8 GB storage.

Setting up

1. Plug one end of your HDMI into the back of the TV.
2. Plug the other end into your device.
3. Plug the power adaptor into your device. Plug the other in the wall.
4. *if using a Ethernet wire plug it into your modem, then the other into your device
5. A light will come on indicating power is being received.
6. Power on your TV. If your TV shows the Amazon Fire TV logo, then your have everything hooked up ok. If not, check to make sure your input is set to the correct HDMI source.
7. It will search for your remote, so make sure the batteries are in it ready to go.
8. When it finishes it will say press **Play** or **Pause** to start.
9. The system will update the software.
10. It will play an introductory video.

Remote

The remote is very easy to use with voice a voice search button build in. You can also download the Fire TV Remote app and use your smartphone.

G-Box Q & Midnight MX2 XBMC

Price and picture

$88

G-Box Q with Kodi Media Center $109

Advantages: Flexibility and customization, Speed
Disadvantages: High cost, Complexity of installations

Content

Netflix, Hulu Plus, Crackle, YouTube, Pandora, TuneIn, Google Play, CNN, Google, Chrome, Firefox, Twitter, Facebook, Skype, Yahoo, ESPN Scorecenter, PlutoTV, HBOGo,

ShowtimeAnytime, WWE Network, Disney Movies, The CW, Watch ABC, FOXNow, StarzPlay, M-GO, EPIX, RabbitTV, entire Google Play App selection.

Speed

The MX2 has a 1.6 GHz Dual core and 1 GB RAM, with 500 MHz quad core GPU. It also has an 8 GB storage Flash.

The Q has a Amlogic Quad Core and 2 GB RAM, with a Octocore GPU. It has 16 GB storage Flash.

Setting up

1. Hook up the HDMI or Composite Video cable(s) to the TV. Connect the other end to the device.
2. Connect the power supply to the device, then plug the other end into the wall or power grid. The device will boot automatically once power is connected.
3. Ethernet connection may be needed HD streaming capability if buffering becomes problematic. Wi-Fi can be connected if you have fast internet.
4. Most customizer users prefer the standard **Launcher** when using the device, over the **3D** or **XBMC** launchers. The standard looks similar to any Android Tablet or Smartphone launcher, so most are already familiar with that UI. It will allow the easiest adding of Google apps from the Play Store.
5. First configure your network connection. To get there click on the 6 dots on top right of the screen (app list button). Click on **Settings**.
   a. If you are going to use Wi-Fi. Slide Wi-Fi slider **ON.** Select your access point, then enter the network password.
   b. For Ethernet connection, enable the Ethernet mode by sliding the Ethernet slider **ON**.
6. You may be prompted to install new updates. If so, do so.
7. You can then run **XBMC** media center and try some of the add-ons. For more information consult XBMC websites and tutorials.
8. You can also run the **Android apps** such as Netflix or download more from the Play store. Consult individual app store information for further details. Also check out the App lists later in the Android device section of this book for a list of specific Android Apps that can be used for your TV.

The Most popular add-ons for the MX2 are Fusion and Xfinity Installers, Mash Up, Ice Films, Navi X, 1 Channel.

Roku 1, 2, 3, LT, Streaming stick
https://www.roku.com/

Roku Streaming Stick $49.99, Roku1 $49.99, Roku2 $64.95, Roku3 $93.69

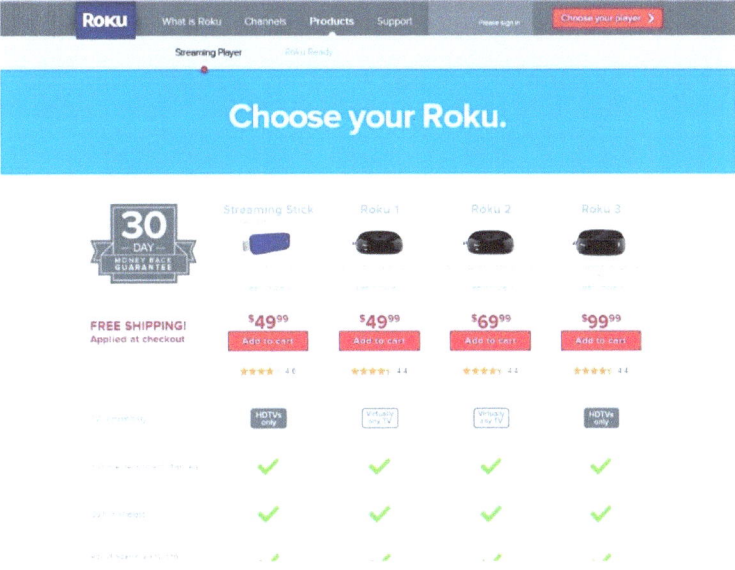

Advantages: Content king, best overall, lower price for slower
Disadvantages: High cost for fastest. RK1 IR remote

Content

Roku sports 1,000+ channels in the streaming sense. Most are on-demand. Some are live streaming. Netflix, Hulu Plus, Amazon Instant, GooglePlay, YouTube, TubiTV, Livestream, VEVO, Crackle, PopcornFlix, Pandora, TuneIn, iHeart Radio, Amazon Cloud, Slacker, AccuRadio, PLEX, VUDU, EPIX, MGO, HBOGO, Showtime Anytime, Weather Underground, Weather Nation, Watch ESPN, MLB.TV, WWE Network, NBA Game Time, NHL Game Center, MLS Live, UFCTV

Speed

Roku 3 has a 900 MHz CPU and is 5 times faster than the Roku 2. **The Streaming Stick, Roku 1, and Roku 2** have 600 MHz CPUs. The Roku 2 varies from the Roku 1 in that the remote is Wi-Fi and the wireless is dual band. So if you get the Roku 1 you may want to connect via Ethernet for best speed since it lacks the dual band Wi-Fi (that is if you have a dual band modem).

Setting up Roku systems

1. Connect HDMI to TV (or composite (SD) for older TVs for Roku 2 if needed).
2. Connect the other end to you Roku Player.
3. Put batteries in the remote (except the Ready-to-go Streaming Stick).

4. Connect power adapter to the Roku, then the other end to the wall.
5. Turn on your TV. If it is in there is no Roku logo, you will need to change the input source to the proper HDMI input mode.
6. Enter your network method. Wi-Fi or Ethernet wire.
7. Roku will tell you to get on to a computer and go to roku.com/link
8. Go there and enter the code shown on the TV on the computer.

Remotes

Roku3 has a Wi-Fi & motion sensor remote. Roku 2 has Wi-Fi. The Streaming Stick (HDMI version) has a RF one. Roku1 has an IR remote (line of sight required).

WD TV Live

Price and picture]

WD TV Live $139.97

WD TV Live Plus $149.98

WD TV Play $91.97

WD TV Media Player $84.93 (lacks Netflix)

http://www.wdc.com/en/products/products.aspx?id=1270

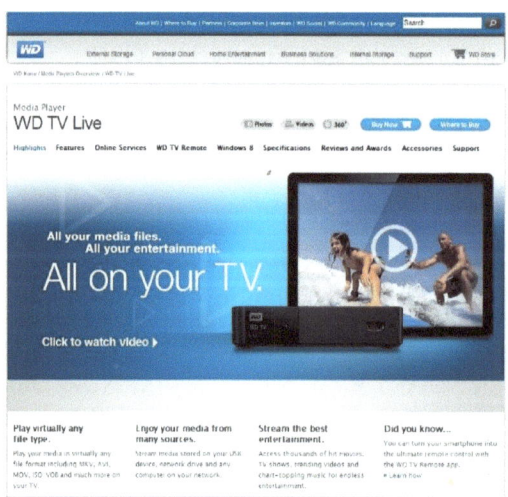

Advantages: Integration with hard drive content, covers major content
Disadvantages: older, High cost, lacks Amazon, lacks sports content

Contents

Netflix, Hulu Plus, YouTube, Pandora, TuneIn, Spotify, Daillymotion, Vimeo, VUDU, Cinema Now, Facebook, AccuWeather, MLB.TV Premium, SnagFilms, RedBull TV

Speed

700 MHz CPU and 512 RAM, 256 Flash Memory.

Year

2011

Setting up

1. Connect the included composite cable up to the TV. Or connect an HDMI cable (not include, but recommended over composite for newer TVs) up to the TV.
2. Connect the other end to the device in the proper port.
3. Plug in the other end to an outlet or surge protector. Connect the other end to the device.
4. Connect a USB storage device to the device if you will be using an external drive.
5. Press the power button on the remote. The indicator light will come on.
6. Turn on your TV. If you are not in the correct source, change the TV mode to the proper input source.
7. Select your language.
A. Connect your device to your network via Ethernet for wire connection. The wizard will begin the automatic connection. When finished press **OK**.
B. For wireless press up and down to find your access point. Press **OK** to select your service. Use the arrow direction buttons to type in your password. Then hit the submit screen button when finished. Press on the **OK** remote button to continue.

Apple TV

https://www.apple.com/appletv/

Price and picture

Model MD199LL/A $89.99

3rd Gen $96.99

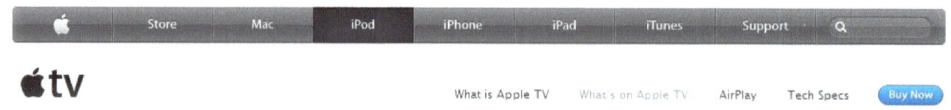

Advantages: Only iTunes content specialization, has major content
Disadvantages: High cost, Lacks Amazon, Google Play

Content

Netflix, Hulu Plus, Crackle, YouTube, VEVO, Vimeo, HBO GO, Showtime Anytime, FX Now, A&E, Lifetime, FYI, History, Watch ABC, Fox Now, Watch Disney, Smithsonian, Red Bull TV, The Weather Channel App, Watch ESPN, MLB.TV, NFL Now, NBA Game Time, NHL Game Center, MLS Live, UFCTV, WWE Network

Speed and resolution

Apple TV 2nd gen 750 MHz and 256 RAM. 720P. **Apple TV 3rd gen** 1 GHz and 512 RAM, 400 MHz GPU. 1080p.

Setting up

1. Connect a HDMI (not included) cable to your TV. Connect the other end to your device.
2. Connect the power supply to the wall. Connect the other end to your device.
3. Turn your TV on. If you don't see the Apple information, make sure your TV is set to the proper input source mode.

4. You should see the Apple logo and language selection screen. Use the direction pad to select the proper language, press the center button to select that language.
5. It will then go to your Wi-Fi setup. Find your access, and enter the password.
6. Choose whether to help Apple with information or not.
7. To use Apple TV with your other computer iTune content go to **Computer** to set up *Home Sharing* option

NETGEAR Neo TV, Neo Max

http://www.netgear.com/landing/stream/tv/#neotv

and pictures

NETGEAR Neo TV (NTV300) (2012) $35.66

NETGEAR Neo TV MAX (2012) $39.99

NETGEAR Neo TV Prime (2012) $65.99

NETGEAR Neo TV Live (2011) $79.99

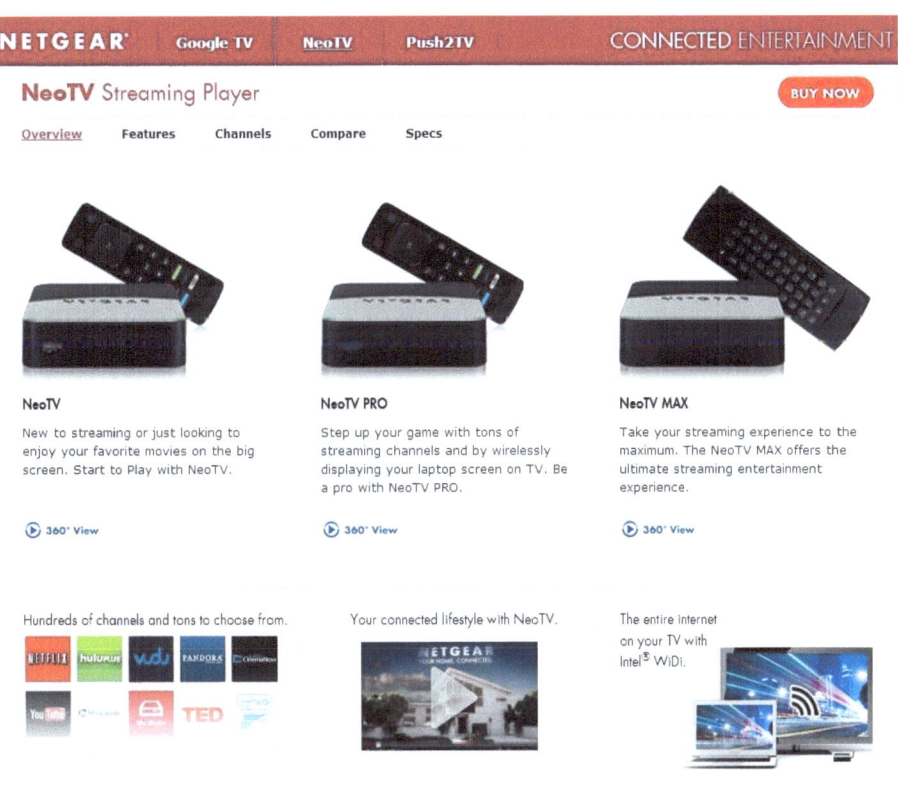

Advantages: lower cost, covers basic streaming
Disadvantages: Lacks major Amazon, iTunes,

Contents

Netflix, Hulu Plus, YouTube, Pandora, Rhapsody, VUDU, Cinema Now, TuneIn, Spotify, TED, TV Guide

Speed

1080p NeoTV Max Mediatek MT8653 CPU and 512 RAM, 256 Mb Flash

Remotes

The Prime and Max remotes have QWERTY keyboards. The Prime model has a Bluetooth remote. The rest are IR remotes.

D-Link MovieNite & Movie Nite Plus

http://us.dlink.com/product-category/home-solutions/entertain/media-players/

Price and picture

$29 MovieNite DSM-310

$39 MovieNite Plus DSM-312

> Advantages: Low cost, basic streaming services
> Disadvantages: Limited content, Lacks most major content providers

Content

Netflix, YouTube, Vudu, Pandora, Picasa

Setting it up

I won't go into details. These are less popular systems. It is basically the same network info and account info as the others. So see the other device setup if help is needed.

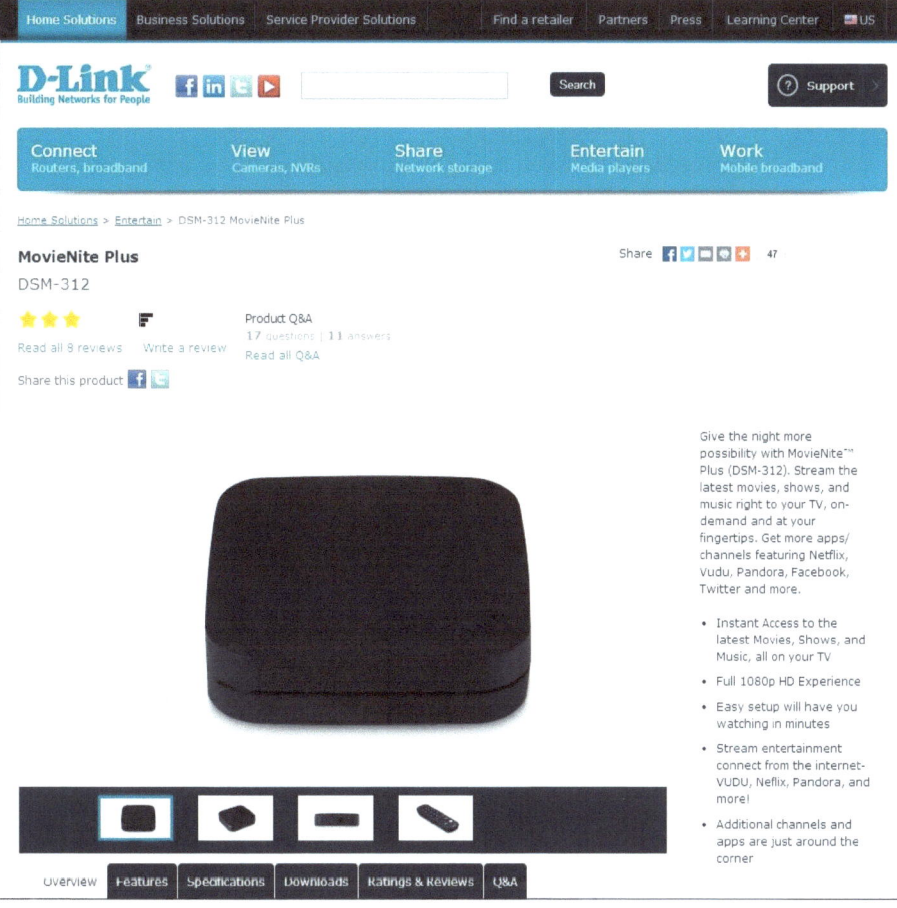

RCA Wi-Fi Streaming Media Player

Price and Picture

$59.60 DSB876WU-WH

$99.99 DSB772E

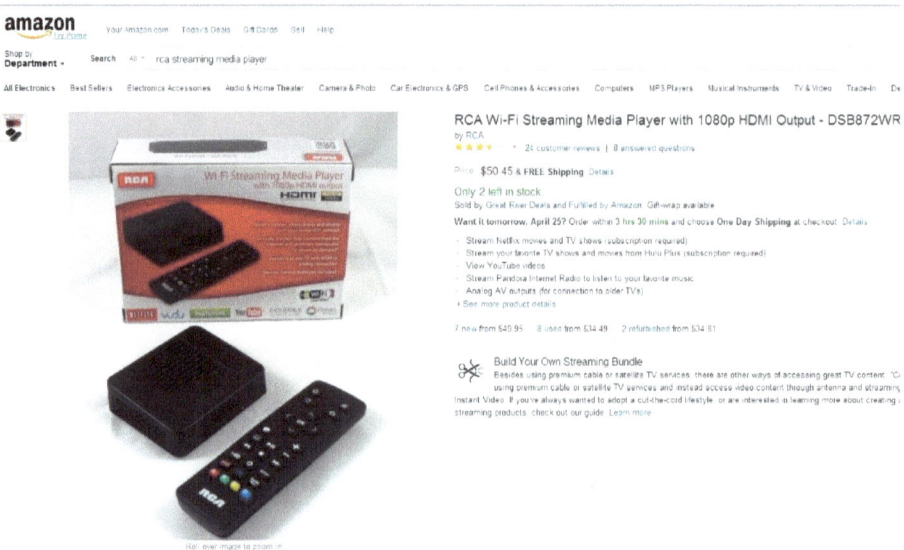

Advantages: basic streaming
Disadvantages: this is a poor streaming at too high price, get something else

Content

Netflix, Hulu Plus, YouTube, VUDU, Pandora, Picasso

VIZIO Co-Star and Co-Star LT

Price and pictures

$89.99Co-Star LT

$99 Co-Star

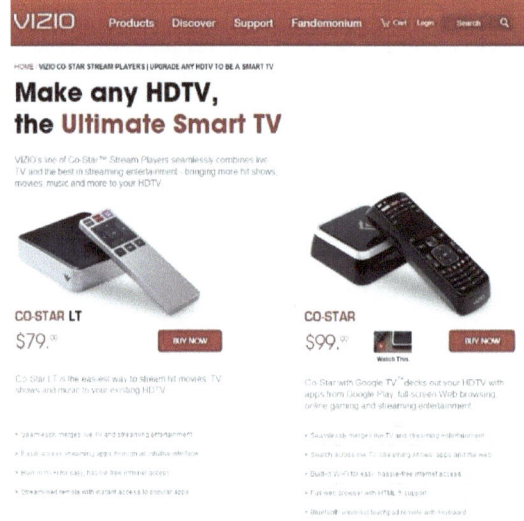

Advantages: integration of Pay TV and hard drive, mid level content availability, mid level speed, full keyboard on Co-star (not LT)

Disadvantages: cord cutters don't need the pay TV integration

Content

This is made to set up with cable or satellite pay TV. So, it will not integrate as it does with cable/satellite sadly with the free OTA TV antenna service. It will still function as a streamer however.

Netflix, Hulu Plus, Amazon Instant, Google Play, YouTube, Crackle, VUDU, MGO, Pandora, iHeart Radio

On live app turns the system into a video game console with Batman Arkham, Lego Hobbit, Lego Movie, Mortal Combat, Metro, and Darksiders II

Co-Star has Chrome Web Browser, LT doesn't.

Speed

1.2 GHz dual-core processor, with a 750 MHz GPU. 1 GB RAM. 4 GB Flash.

Remote

LT is IR. Co-Star has Bluetooth remote with touchpad and built in keyboard on backside.

Android Mini PCs

Popular the last few years is the Android Mini PCs. These devices essentially turn your TVs into giant tablet computers without the touch screens.

Price and picture:

Advantages: Flexibility and customization, Speed, Low cost

Disadvantages: Complexity of installations, no remote (though smartphone app remote)

What to look for in specifications

Android OS

Newest 4.4 **KitKat** – User interface appearance changes, optimized for low RAM devices, wireless printing, select default text messaging and home launcher app. 5.3% Android devices. For more features see: http://developer.android.com/about/versions/kitkat.html

Most Popular 4.1 to 4.3 **Jelly Bean**

4.3 - AV Bluetooth remote support, OpenGL ES 3.0 support. 8.9% Android devices. For more features see: http://developer.android.com/about/versions/android-4.3.html

4.2 - Lock screen improvement, "Daydream" screen saver when docked, Miracast support, blind user support, new clock/timer, group SMS, Bluetooth gamepads and joystick support (4.2.1), 18.1% Android devices. For more features see: http://developer.android.com/about/versions/android-4.2.html

4.1– Smoother UI, bi-directional text (for multi languages), user installed keyboard maps, turn off notifications per app, rearrange + resize widgets, Bluetooth data transfer Android Beam, offline voice dictation, better camera, better voice search, USB audio supported. 34.4% Android devices. For more features see: http://developer.android.com/about/versions/android-4.1.html

4.0 **Ice Cream** - Visual voicemail, screen capture (power button + volume-down button), lock screen app access, better copy/paste, continuous voice dictation, face unlock, shutdown background data apps, improved camera, new gallery layout, new photo editor, Wi-Fi Direct, Android Beam near field communication. 14.3% Android devices. For more features see: http://developer.android.com/about/versions/android-4.0-highlights.html

3.0 **Honeycomb** – New UI, simplified multitasking w/ snapshot, new keyboard, simplified copy/paste, multi browser tabs, multi core processor support, encrypted data, USB accessory connection, external keyboard and mouse support, joystick and gamepad support. For more features and details see: http://developer.android.com/about/versions/android-3.0-highlights.html

2.3 **Gingerbread** – VoIP support, copy/past improvement, NFC, Plus all the updates from the prior versions which will not be shown here. 17.8% Android devices. For more features see: http://developer.android.com/about/versions/android-2.3-highlights.html

Android statistics are from developer.android.com (Android OS, 2014)

CPUs & GPUs
The CPU performs most app tasks and processing. The GPU performs the graphic processing. More cores allow more multi-tasking of processing. Bolded is the top current processers.

ARM Cortex A8 -OMAP 3 600 MHz mostly to 1.2 GHz, GPU PowerVR SGX530 – 277 to 384 MHz
Snapdragon S1 – 528 MHz to 1 GHz, GPU Adreno 200 – 200 MHz
Snapdragon S2 – 800 MHz to 1.5 GHz Dual Core, GPU Adreno 205 – 333 MHz
Snapdragon S3 – 1.7 GHz Dual Core, GPU Adreno 220 – 500 MHz

Snapdragon S4 – 1 to 1.7 GHz Dual and Quad Core, Adreno 203, 225, 305, & 320 - 500 to 533 MHz

Snapdragon 200 – 1.2 to 1.4 GHz Dual and Quad Core, Adreno 203 & 302

Snapdragon 400 – 1.2 to 1.4 GHz Dual and Quad Core, Adreno 305 – 450 to 533 MHz

Snapdragon 600 – 1.5 to 1.7 GHz Quad Core, GPU Adreno 320 – 400 MHz

Snapdragon 800 – 2.26 GHz Quad Core, GPU Adreno 330 - 450 MHz

Exynos 3 – 1 to 1.2 GHz, GPU IT PowerVR SGX540 - 200 MHz

Exynos 4 – 1.2 to 1.6 GHz, Dual and Quad Core, GPU ARM Mali-400 MP4 - 266 to 522 MHz

Exynos 5 – 1.7 to 2.1 GHz, Dual and Quad Core, GPU ARM Mali-T628 & 624 MP6 or IT PowerVR SGX544MP3 - 480 to 695 MHz

Exynos 5420 - 1.9 GHz octa-core, a six-core ARM Mali-T628 GPU 695 MHz

Tegra 2 – 1 to 1.2 GHz, Dual Core, GPU 300 to 400 MHz

Tegra 3 – 1.2 to 1.6 GHz, Quad Core, GPU 416 to 520 MHz

Tegra 4 – 1.9 GHz, Quad Core, GPU 672 MHz

RAM

RAM is the working memory of the computer device.

256, 512, 768, 1024 (1GB), 2048(2GB)

Internal Storage

16GB, 32GB

Micro SD Card Slot

Varies

Connectivity

Wi-Fi 802.11 a/b/g/n/ac, Bluetooth 2.0 + EDR, Bluetooth v2.1 with A2DP, Bluetooth 3.0, Bluetooth v4.0 HS, A-GPS,

Input

Micro-USB; USB On-the-go, 3.5 mm

Output

Screen
HDMI
Micro HDMI
Mini HDMI

Streaming Apps for Mobile and Mini Devices

Android Apps

Network

NBC ABC FOX (CBS, CW) – see Hulu Plus app Hulu Plus. Android Varies

https://play.google.com/store/apps/details?id=com.hulu.plus

CBS http://www.cbs.com/mobile/ Android 4.0 and above. Episodes

CW https://play.google.com/store/apps/details?id=com.cw.fullepisodes.android Android 2.2 and above. Episodes

ION https://play.google.com/store/apps/details?id=com.iontelevision.mobile . Clips. Android 4.0 and above

Public TV

PBS MHz Network https://play.google.com/store/apps/details?id=com.mhz.MhzNetworks Android 1.6 and above. Live, Episodes

PBS NHK World (Japan) https://play.google.com/store/apps/details?id=jp.nhkworldtv.android Android 2.2 and above. Live

PBS France 24 https://play.google.com/store/apps/details?id=com.france24.androidapp Live and on demand. Android varies

Music TV

Zuus https://play.google.com/store/apps/details?id=com.zuus.app Android 2.2 and above. 50 music genre stations.

Weather

Accuweather https://play.google.com/store/apps/details?id=com.accuweather.android Android 2.2 and above

Weather Nation https://play.google.com/store/apps/details?id=com.xav.wn Android 2.3 and above

Streaming TV

Hulu Plus https://play.google.com/store/apps/details?id=com.hulu.plus

Streaming movies Android

Netflix (subscription required)

https://play.google.com/store/apps/details?id=com.netflix.mediaclient Android Varies

Crackle (free movies) https://play.google.com/store/apps/details?id=com.gotv.crackle.handset Android 2.3.3 and above

VUDU (Pay per)

https://play.google.com/store/apps/details?id=air.com.vudu.air.DownloaderTablet

PRICE COMPARISON CHART 2015

System	System cost (minus other costs)		
Existing PC, Tablet, notebook	Free		
Existing Wii or Playstation	Free		
Existing Xbox	Gold membership $50 annually		
Existing SmartTV	Free		
Existing Wi-Fi Blue-ray Player	Free		
D-Link MovieNite	$29.00		
Chromecast	$31.59		
NETGEAR NeoTV	$35.66		
Amazon Fire Stick	$39.00		
D-Link MovieNite Plus	$39.00		
NETGEAR NeoTV Max	$39.99		
Roku Streaming Stick	$49.00		
Roku1	$49.99		
Older Blue-ray Streaming Player	$50.00		
Older Android Mini PC	$50.00		
Roku2	$64.95		
NETGEAR NeoTV Prime	$65.99		
RCA Streaming Player	$67.95		
NETGEAR NeoTV Live	$79.99		
Newer Android Mini PC	$80.00		
WD TV Media Player	$84.93		
G-Box MX 2	$88.00		
Apple TV	$89.99		
Vizio Co-Star LT	$89.99		
WD TV Play	$91.97		
Roku3	$93.69		
Apple3 TV	$96.99		
Vizio Co-Star	$99.00		
Amazon Fire TV	$99.00		
RCA Streaming Player DSB772E	$99.99		
Mid Grade Blue-ray Streaming Player	$100.00		
Wii	$100.00		
G-Box Q	$109.00		
WD TV Live	$139.99		
WD TV Live Plus	$149.97		
Xbox 360	$163.00		
Playstation3	$200.00		
Low price small SmartTV (various)	$200.00		
New WiFi Blueray (various)	$200.00		
PS Vita	$200.00		
Wii U	$250.00		
Xbox One	$350.00		
PS 4	$399.00		

	Year	Price	Installation	Content	Remote Control	Mobile Remote Apps	Speed	Memory	Blue-ray and DVD
Xbox	360 2005, One 2013	$179 to $500 XBoxes; $50 annual Live Gold membership	$50 annual fee a minus	Great	$20 separate unless comes with bundle	NA	360 3.2 Ghz Triple, One 1.7 8core	4GB to 160GB (XBOne 500GB)	Yes
Wii	Wii 2006, Wii U 2012	$100 to $400	EZ	Limited	controller	NA	Wii 729Mhz, Wii U 1.24 Ghz 3core	512MB + SD card (8 to 32GB Flash)	No
Playstation	PS3 2006, PS4 2013	$200 to $600	Ez	Great	$10 to $40	NA	PS3 3.2 Ghz, PS4 1.6 Ghz 8core, PSVita 2Ghz 4core	20 to 160GB (PS4 500GB), PS3 256 Mb or PS4 + Vita 512 Mb RAM	Yes
Smart TV (various)	Samsung 2012 VIZIO 2012 - 2013 LG 2013 SONY 2013	$200 +	EZ	Good	in TV remote	Sony, LG, VIZIO, and Samsung	2yrs old	Normally under 1GB	If built in
WiFi Blueray (various)	2013	$50 to $200	EZ	Good	Yes	Sony and Samsung	1yr old	Normally under 1GB	Yes
Chromecast	2013	$35	Med	Ok	laptop, tablet, or smartphone	iOS and Android Apps, Chrome Browser	1.2 Ghz	512 Mb	No
Amazon Fire TV	2014	$99	EZ	Good	Yes, with voice search	Kindle Fire	1.7 Ghz quad	2 Gb RAM, 8 Gb	No
G-Box Midnight MX2 XBMC	2013	$98	Complex	Depends	Yes	iOS, Android, Kindle Fire, Windows, Windowsphone	1.6 GHz Dual core	1 GB RAM, 8 Gb Flash	No
WD TV Live	2011	$84.99	Complex	Ok	Yes	iOS and Android Apps	700 MHz CPU	512 RAM	No
Roku	Roku 1, 2, 3, LT 2013; Roku Streaming Stick 2014	$49.99 to $94.99	EZ	Great	Yes		RK3 900 Mhz, RK2 + RK1 600 Mhz	256Mb	No
Apple TV	Apple TV 2012; Apple TV3 2014	$92.95 to $114.99	Med	Good	Yes	iTunes and iOS	2nd Gen 750 Mhz, 3rd Gen 1 Ghz	2nd Gen 256 RAM, 3rd Gen 512 RAM	No
NETGEAR NeoTV	Neo TV 2011; Neo TV Max 2012	$43.99 to $74.85	EZ	Ok	Yes	iOS and Android	Slow	Neo Max 512 Mb RAM, 256 Mb Flash	No
D-Link MovieNite Plus	2012	$40.47	EZ	Limited	Yes	iOS and Android	Slow	?	No
RCA Streaming Player	2012	$59.47	Poor	Limited	Yes	no	Crawling	?	No
VIZIO Co-Star	Co-Star 2012 Co-Star LT 2013	$58 to $86	Med	Good	Yes	no	1.2 Ghz Dual core	1 Gb RAM, 4 Gb Flash	No
Android Mini PC	RK 3188	$50 to $80	Complex	Depends	No	Android	1.6 Quad core	2 Gb RAM, 8 Gb Flash	No

Streaming Sources

1. Antenna + Hulu free or Hulu Plus subscription + news network website clips, articles, and episodes = rough equivalent to basic cable.

2. Get a Library Card and check out what movies they have and what online streaming services they might have. My library has free FilmsOnDemand and Indieflix, free music, and 50 free magazines from Zinio.

3. Add free Crackle, Popcornflix, Livestream, PlutoTV, TubiTV, SnagFilms, Filmon, and YouTube content.

4. Add to the above an EPIX movie subscription (all have a free trial month) of either Amazon Prime ($99/yr) or Netflix ($7.99) depending on your bonus focus. Amazon Prime includes free 2 day shipping with amazon purchases, Kindle Lending Library, A&E and History channel series like Duck Dynasty and Ancient Aliens. Netflix includes older TV series, foreign, and older movies in addition to the EPIX current first run movies. **Redbox kiosks offers release from their Kiosk $1.50.** You might even want to rotate through the different services though Amazon Prime a year once the trial is completed.

5. Premium Series not included in the options above you can buy and stream from Amazon, iTunes, VUDU, YouTube, Google Play, CinamaNow, or M-Go normally for SD 1.99 to $2.99/episode or HD 3.99 to 4.99/episode. Only do this for content not included in the free or subscription services above.

6. Add sports entertainment package if you desire. NFL, MLB, NBA, NHL, WWE. Most include streaming anywhere.

7. Rent Movies subscription Netflix Disk, Redbox Kiosk, or on a per use basis Amazon Instant, iTunes, Google Play, VUDU, CinamaNow, Blockbuster On Demand

8. Prices are determined by what content you enjoy.

Hulu streaming plus CBS through their TV.com streaming services really stands alone together for Network Broadcast TV shows recently played. They also play a few entire series, many times requiring Hulu Plus $7.99 per month. In addition, Hulu also has a lot of basic cable content.

My Recommendations for streaming services

The main consideration for which services to subscribe to is HuluPlus and which other streaming service subscription should you consider.

Hulu Plus, although it has some movies, really is about TV Shows. There are a few movies.

Largest amount of content: If you are looking for the largest amount of older films and TV shows then Netflix is the best.

If you are looking for new releases: Then Redbox kiosk with their low price kiosk rentals is the best deal. Netflix has their new releases with their DVD plans, though waiting for them can be frustrating sometimes some new releases taking forever to be shipped, and you need to subscribe to their disk mail services. For Redbox, you must live near a kiosk, so check on their website for available content. Even then, Netflix has long waiting lists many times for new release disks.

If you read ebooks and order from Amazon.com: Amazon Prime may be the streaming service for you. In addition, you gain free book checkout service from their Lending library, and free shipping on qualified orders.

Free route: Hulu free (on computer), with library card content, with Crackle, Snag films, PopcornFlix, PlutoTV, TubiTV, YouTube movies, free trial offers, & Filmon movies. Also you can use the list at the end of the book to find free streaming and episodes from several networks.

Streaming Service Providers

Hulu (NBC, ABC, Fox, the CW, ION, WB, BBC, MTV, BET, VH1, Nat Geo, Syfy, Spike, TBS, TNT, USA, Oxygen, Lifetime, LMN, E!, AMC, WWE, Food Network, Disney, Comedy Central, A&E, Bio, DIY, HGTV, History, H2, TVLand, Travel Chan, IFC, Jim Henson, Cartoon Network, Adult Swim, Nickelodeon, PBS Kids, Anchor Bay, Image Ent, Lions Gate, MGM, Anime Net, Aniplex, Funimation, Bandai, Manga, TokyoPop, TOEI, VIZ, Exercise TV, GAIAM TV, Fora TV
http://www.hulu.com/

Hulu has two services 1) free and 2) Plus. Most devices only can receive Hulu Plus. Hulu Plus currently is a free week followed by $7.99

Hulu is a joint venture of NBC Universal Television Group (Comcast), Fox Broadcasting Company (21st Century Fox) and Disney–ABC Television Group (The Walt Disney Company), with funding by Providence Equity Partners, the owner of Newport Television

Movie and TV Library

FOX: FOX, FX, Fox News Channel, Fox Business Network, National Geographic Channel
NBC Universal: NBC, CNBC, MSNBC, Syfy, Oxygen
Walt Disney Company: ABC, ABC Family
A&E Television Networks: The Biography Channel, History Channel, A&E
Turner Broadcasting System/Warner Brothers: TBS, TNT, Cartoon Network, truTV, CNN, Adult Swim, The CW
AMC Networks
ION Television but not the Qubo shows
Food Network: Cooking Channel
WWE shows
Anchor Bay

Additionally, Hulu offers CBS Broadcasting content in Japan and now links the US Original Content

Hulu Original Series
A Day in the Life (2011–2012)
Battleground (2012)
Spoilers (2012)
Up to Speed (2012)
The Awesomes (2013)
Quick Draw (2013)
The Wrong Mans (2013)
Behind the Mask (2013-)
The Hotwives of Orlando (2014)

Popular TV Shows
Modern Family
Law and Order: SVU
South Park
Nashville
New Girl
The Daily Show with John Stewart
Family Guy
The Tonight Show with Jimmy Fallon
Revolution
Naruto
Chicago PD
The Voice
Grey's Anatomy
SpongeBob Squarepants
Glee
Suburgatory
American Dad
Community
The Middle
Bones
The Originals
The Goodwife
Parks and Recreation

Chicago Fire

Trophy Wife

Once Upon A Time

The Goldbergs

Scandal

The Following

Dragonball Z

Bachelor

Arrow

Saterday Night Live

Raising Hope

One Piece

Pokemon

Castle

Sabrina

Late Night with Seth Meyers

The Vampire Diaries

Doctor Who

Jimmy Kimmel Live

Cosmos

Popular Movies

The Usual Suspects

Good Dick

Bad Kids Go to Hell

Steel Magnolias

Lonely Hearts

Husbands and Wives

Babe

Veggie Tales

Resident Evil: Apocalypse

The Girl with the Dragon Tattoo

Dark House

Who's Harry Crumb

Akira

Pokeman

Brutal

Atlantic Rim

Intolerable Cruelty

Lost in Translation

CBS All Access

www.cbs.com/all-access/

This is a new service offered by CBS for the price of $6 a month. The service includes next day aired shows, less ads [though not ad free on new shows], and ad free classics archive.

CBS Content

CSI: Crime Scene Investigation
NCIS
Criminal Minds
The Mentalist
NCIS: Los Angeles
The Good Wife
Hawaii Five-O
Blue Bloods
Person of Interest
Elementary
Under the Dome
Extant
Madam Secretary
Scorpion
NCIS: New Orleans
Stalker
Two and a Half Men
The Big Bang Theory
Mike and Molly
2 Broke Girls
Mom
The Millers
The McCarthys
Big Brother
Survivor
The Amazin Race
Undercover Boss
Late Show with David Letterman
The Late Late Show with Craig Ferguson
The Young and the Restless
The Bold and the Beautiful
The Talk
Let's Make a Deal
The Price is Right
60 Minutes
48 forty eight hours

Classic CBS

The Andy Griffith Show
Beverly Hills 90210
Brady Bunch
Cheers
CSI: Miami
Everybody Loves Raymond
Family Ties
Frasier
Ghost Whisperer
I Love Lucy
Macgyver
Medium
Melrose Place
Mission Impossible
Perry Mason
Sabrina the teenage Witch
Star Trek [All series]
Taxi
Twilight Zone [original]
Twin Peaks
Wings

Netflix (EPIX) (Back catalog Time Warner, Universal, MGM, Paramount, Lions Gate, Sony, 20th Cent Fox, Disney)
https://www.netflix.com/

Netflix's "Watch Instantly" service holds first-run rights to films from Paramount Pictures, MGM, Lions Gate Entertainment (through an output deal with Epix), along with back-catalog titles to films from Time Warner, Universal Pictures, Sony Pictures, Paramount Pictures, MGM, Lions Gate Entertainment, 20th Century Fox, Disney, and other distributors. Netflix also provides current and back-catalog TV programs distributed by NBC Universal, 20th Century Fox, Sony Pictures, Disney-ABC Domestic Television, with select shows from Warner Bros. as well. Netflix has "pay TV window" deals with Relativity Media, FilmDistrict, and Open Road Films.

Original Content
House of Cards,
Hemlock Grove, and
Orange Is the New Black
Arrested Development

Netflix received a five-season deal for four Marvel Super Heroes: Daredevil, Jessica Jones, Iron Fist, and Luke Cage. The deal involves the broadcast of four 13-episode seasons that culminate in a mini-series called The Defenders. Broadcasting will commence in 2015.

Netflix Instant Video
Major Studio New Movies (Based on current contracts)

EPIX
Paramount
For a list of recent movies by Paramount
http://en.wikipedia.org/wiki/List_of_Paramount_Pictures_films#2010s

MGM
For a list of recent movies by MGM
http://en.wikipedia.org/wiki/List_of_Metro-Goldwyn-Mayer_films#2010s

Lionsgate
For a list of Lionsgate movies
http://en.wikipedia.org/wiki/List_of_theatrically_released_Lionsgate_films#2010s

Relativity Media
For a list of Releativity movies
http://en.wikipedia.org/wiki/Relativity_Media

FilmDistrict
For a list of FilmDistrict movies
http://en.wikipedia.org/wiki/FilmDistrict

Open Road Films
For a list of Open Road Films movies
http://en.wikipedia.org/wiki/Open_Road_Films

New and Old TV shows (based on major studio contracts)
NBC http://en.wikipedia.org/wiki/Universal_Television
20th Century Fox http://en.wikipedia.org/wiki/20th_Century_Fox_Television
Sony Pictures http://en.wikipedia.org/wiki/Sony_Pictures_Television
Disney-ABC http://en.wikipedia.org/wiki/Disney%E2%80%93ABC_Domestic_Television
Warner Brothers (select shows) http://en.wikipedia.org/wiki/Warner_Bros._Television

Popular TV shows

New Girl FOX– past seasons when season is over

The Walking Dead AMC – past seasons when season is over

Mad Men AMC – past seasons when season is over

How I met your Mother CBS– past seasons when season is over

Family Guy FOX– past seasons when season is over

Archer FX – past seasons when season is over

Clone Wars TOON – current series

Good Luck Charlie DISN – past seasons when season is over

Completed Series

Breaking Bad

Dexter

Amazon Instant Video (EPIX)

Amazon Instant Video is an Internet video on demand service offered by Amazon in the United States, United Kingdom, Germany and Japan. It offers television shows and films for rental and purchase. Except in Japan, a number of titles are available free through to customers with an Amazon Prime subscription.

Amazon instant videos rented, bought, and free through Amazon Prime can download to computers, a few media boxes like TiVo and Roku, a few smart TVs, console gaming units such as PS3, Xbox, and Wii, Apple IPhone and IPad, and Kindle Fire. Check with device specification features to see if Amazon instant videos will work with them.

Amazon Instant Video Membership

Amazon Instant Videos can be purchased or rented on an individual basis. Also Amazon has a Prime membership that includes some free movies, TV shows, Kindle Owner's lending library, and includes some free two day shipping on some qualified items bought at Amazon.com. Currently, Amazon Prime offers a free month trial, and $99 a year.

Movie and Video Content

EPIX (pronounced "epics") is an American hybrid premium cable and satellite television network, and subscription video on demand service that is operated by Studio 3 Partners LLC, a joint venture of Viacom (specifically its subdivision Paramount Pictures), Metro-Goldwyn-Mayer and Lions Gate Entertainment.

HBO and Showtime runs older series now on Amazon available through prime.

Amazon Prime Popular Movies

Star Trek: Into Darkness

The Hunger Games

Marvel's The Avengers

Frozen Ground

The Spectacular Now

Pain & Gain

Jack Reacher

Robin Hood Men in Tights

The Guilt Trip

Stardust

Clerks

Skyfall

Hansel and Gretel

Flyboys

Goonies

Temptation

Flight

Miss Congeniality

Spaceballs

Enigma

Eden Log

All Good Things

House

Grimm

Beetlejuice

Caddyshack

Star Trek: Wrath of Khan

Popular TV shows

The Americans

The After

Vikings

Downton Abbey

Orphan Black

Hannibal

Veronica Mars

HBO
Sapranos
True Blood
The Wire
Boardwalk Empire
Band of Brothers
Deadwood
Eastbound & Down
Six feet under
Treme
Enlightened
Oz

Original Content
Amazon premiered the comedies Alpha House and Betas.

Redbox Instant (EPIX) - Shutdown Oct 10, 2014
See message of shutdown http://about.redboxinstant.com/news/
Redbox Kiosk available titles. http://www.redbox.com/ also has links to iOS and Android apps

HBO Go

HBO is said to be launching a streaming stand alone service in April of 2015. Because of this, I have added HBO to the app comparison chart replacing Redbox Instant subscription. Rumors place it at $15/month, though unconfirmed at the moment. It is scheduled to launch with the season return of Game of Thrones. HBO Go is supported by many devices.

Crackle (Sony)
Crackle is backed by Sony Pictures Entertainment

Original Programs: Shows include "Bannen Way" "Comedians in Cars Getting Coffee" "Chosen", "Cleaners", "Trenches", "Unknown", "Sports Jeopardy", "Sequestered"

Original full feature film; "Extraction"

Movie and TV Library
Crackle features many Columbia Pictures and Screen Gems titles including Men In Black, Pineapple Express, District-9, Step Brothers, Talladega Nights, Blankman, Fletch, Conan.

Crackle features Sony distributed television series like "Damages", "Rescue Me", "The Shield" and "Seinfeld".

Crackle's content refreshes monthly with titles being added and taken down.

Content Partners
Aniplex
FOX Digital
Funimation
Lionsgate
MGM
Red Bull
SnagFilms
TOEI

Genres
Crackle features programming in the following key genres: Action, Comedy, Crime, Drama, Horror and Sci-fi.

In January 2012, Crackle added Animex anime channel to its content, available to viewers in the US and Canada.

Top Titles
Movies
Resident Evil: Afterlife
Sol
Attack the Block
The Roommate
Under Suspicion
Talladega Nights: The Ballad of Ricky Bobby
Slackers
Edison Force
Booty Call
Run Lola Run
Insanitarium
April Fool's Day
Candyman
American Psycho
Zonad
Kaena: The Prophecy
Godzilla 2000

TVShows

Jeopardy! Flashback

Queen's Blade

Marvel Anime: X-Men

The Jackie Chan Adventures

Barney Miller

Aim High

All In The Family

Bewitched

Blood+

The Border

Damages

Good Times

I Dream of Jeannie

Married With Children

News Radio

Rescue Me

YouTube Movies (Google: Merged with Google Video)
https://www.youtube.com/user/movies

You can buy movies, but here is a link to free movies offered on YouTube and a list of the main channels/studios.
Free YouTube
Movieshttp://www.youtube.com/user/movies/videos?sort=dd&shelf_id=12&view=26

Movies by
Starz Media (Anchor Bay) http://www.youtube.com/user/starzmedia
Maverick Movies http://www.youtube.com/user/maverickent
http://maverickmoviesonline.com/
Entertainment One Benelux http://www.youtube.com/user/eonebenelux
Gravitas Ventures http://www.youtube.com/user/GravitasVOD
Cinedigm http://www.youtube.com/user/NewVideoDigital
Abehorror http://www.youtube.com/user/Abehorror
I dream Productions http://www.youtube.com/user/IDreamProduction
Viewster http://www.youtube.com/user/ViewsterTV
FirstLookStudio (Millennium Entertainment) http://www.youtube.com/user/FirstLookStudios
Popcornflix http://www.youtube.com/user/ScreenMediaPictures

Cinama Nirvana http://www.youtube.com/user/CinemaNirvana
Docurama Films http://www.youtube.com/user/docurama
Drelbcom http://www.youtube.com/user/drelbcom
VISOCinema http://www.youtube.com/user/VISOCinema
MANGAEntertainment http://www.youtube.com/user/MANGAentertainment

Popcornflix (Screen Media)
http://www.popcornflix.com/

The site primarily consists of independent feature films, many of which come directly from Screen Media's library. No subscription is required rather it is ad based.

Top Movies
Dead Tone
The Cry
The Toxic Avenger
Class of Nuke em High
Easy
Two Brothers and a Bride
2 Days
Descent
30 Years to Life
Dream Warrior

TV Series
Monster Garage
Cheaters

National Geographic Channel
Life Below Zero
Icons of Power
Tools of the Trade

PlutoTV
http://pluto.tv/#!

A recent start up which takes YouTube content and turns it into more of TV like continuous content, curated by people. There are 100+ channels covering music, news & info, lifestyle, education, sports, comedy, tech, entertainment, art & culture, and kids. Although not new content, it serves as an intermediary and organizes the YouTube content into continuous content streams. It transforms some YouTube content into user friendly streams.

This is fun and works rather well.
Available PC, Mac, Amazon Fire TV, Chromecast, and Apple TV (as of 11/14/2014)

TubiTV
http://tubitv.com

Tubi TV content partners include Starz Digital Media, Cinedigm, Shine International, Jim Henson Co., Hasbro Studios, Film Movement, ITV, Endemol, Zodiak Rights, DRG, All3Media, Kino Lorber, Korean TV network MBC and Korean studio CJ Entertainment. In addition, Tubi TV has lined up several digital content partners, which include Newslook, AP, Reuters, anime distributor Funimation, Havoc Television, ACC Digital Network, Viki, Anyclip.com and Wochit.

It has some unique content not found elsewhere helping it supplement other services nicely.

Available Roku, Amazon Fire TV, Android, iOS

Snag Films
http://www.snagfilms.com/

Documentary and Independent films. National Geographic and PBS also places their produced films here.

Filmon
http://www.filmon.com/

Has live and VOD free content, with occasional commercial interruptions.

Live UK channels, horror, classic tv, some news channels, music, entertainment, gaming, lifestyle, sports, tech

With apps for Roku, PC, Android, Windows mobile, and iOS among others.

VUDU (Walmart)
VUDU is a video on demand (VOD) video delivery company owned by Walmart.

15,000 total titles in the VUDU catalog, including both movies and television shows. VUDU has established content licensing contracts with all major movie studios as well as over 50 smaller and independent studios.

They offer a 5 free video HD streaming movies from select titles.

VUDU can stream videogame consoles, many Blu-ray players with build in smart functions, iOS mobile and Android devices. Roku, Neo, and many streaming media devices.

New Rentals are $3.99 to $5.99 depending on definition.

New Purchases are $14.99 to $19.99 depending on definition.

TV series can be $1.99 to 2.99 per episode.

Target Ticket

Similar to Walmart's VUDU service they have similar prices as well. They offer 10 free video HD streaming video downloads at sign up from a pre-selected pool.

M-GO (Dreamworks and Technicolor)

Offers 2 free movies, after that pay-as-you go.

Partner of Dreamworks and Technicolor.

New Rentals are $3.99 to $4.99 depending on definition.

New Purchases are $14.99 to $19.99 depending on definition.

TV series are $1.99 per episode.

Sony Entertainment Network

Similar to other download movie services in price.

Fandor

Film festival type films. Independent, classic, foreign, and documentary. Available on many devices. Android, iOS, Kindle, Roku, PC & Mac

$90 annually or $10/month

ErosNow

http://erosnow.com/

Hindi movies. Some are free. Some require subscription.

DramaFever

http://www.dramafever.com/

Watch Korean dramas, TV shows, and movies free.

Sports Packages

Make sure to check whether or not your device will be able to use these sports package services.

MLB

http://mlb.mlb.com/mlb/subscriptions/

MLB.TV Premium $99/year, $24.99/month. Desktop & laptop as well as mobile and streaming devices. You get home and away announcer audio feeds, mobile support, and At Bat 14 app for mobile included.

MLB.TV Basic $79.99/year, $19.99/month desktop or laptop only.

MLS

http://www.mlssoccer.com/mls-live

2014 prices - MLS $49.99/season Roku, Apple TV, iPad, iPhone, Android, & Panasonic Devices

NASCAR

http://www.nascar.com/en_us/ajax/static/raceview-product-page.html

2014 prices -

NASCAR Raceview Premium $44.95/season, $9.95/month. PC only. This is 3D virtual models rather than video feed. Statistics

NASCAR Raceview Audio $19.95/season, $4.95/month. PC only Broadcast and Driver audio.

NASCAR Raceview Mobile Premium $39.99/season, 4.99/month. 3D virtual race & audio of drivers and broadcast.

WNBA

www.**wnba**.com/**Live**Access

WNBA Live Access 2014 $9.99/ Season

NFL (Computers and Windows, Android, iOS Tablets)

http://www.nfl.com/qs/GameAccess/index.jsp

NFL Preseason $19.99. Live games except blacked out games, which you can watch 24 hours after the game completes.

NFL Game Rewind Season Plus $59.99/season or 4 payments of $17.99/month. Replays of every NFL Game, Playoffs, and Super Bowl (not live).

NFL Game Rewind Season $29.99. Replay of every NFL game during regular season (not live).

NFL Game Rewind Follow Your Team $24.99/season. Replay of every game of your favorite team during the regular season (not live).

https://audiopass.nfl.com/nflap/secure/packages?ttv=0

NFL Audio Pass Season Plus $19.99/season. Listen to every game live including Playoffs and Super Bowl. Listen to archived games.

NFL Audio Pass Season $14.99/season. Listen to every regular season game live. Listen to archived games.

NFL Audio Pass Follow Your Team $9.99/season. Listen to your favorite team's games live. Listen to archived games.

NHL
www.nhl.com/gcl/

NHL Gamecenter Live 2014-15 $99.95/season, or 5 payments of $19.99/month.

NBA
http://www.nba.com/leaguepass/

NBA League Pass - TV, internet, mobile $199/Season; Digital only- Internet and mobile $149 (although it's half off halfway through this season)

UFC
http://www.ufc.tv/page/fightpass

UFC Fightpass - $9.99/month, 6m commitment $8.99/month, 12m commitment $7.99/m. Live Fight Nights, UFC Unleashed, Best of Pride, The Ultimate Fighter, originals

WWE Network
http://www.wwe.com/wwenetwork

$9.99/month. Includes live all broadcasts streaming, original content, archives, & live pay per views.

Popular Free Sports Online and Apps
NFL Now & NFL Mobile- NFL Network clips, scores, and news   http://now.nfl.com/

Fox Sports - http://www.foxsports.com/

Watch ESPN - http://espn.go.com/watchespn/

Hunting, Outdoors, and Fishing - http://www.carbontv.com/

Huntit.TV http://www.huntit.tv/ hunting, fishing

NETWORK TELEVISION (Online Website Links and Resources)

Commercial Television

*ABC   http://abc.go.com/   watch episodes

ABCNEWS   http://abcnews.go.com/   clips and news

*CBS   http://www.cbs.com/   watch episodes

CBSNEWS   http://www.cbsnews.com/   clips and news

*CW   http://www.cwtv.com/   watch episodes

*FOX   http://www.fox.com/   watch episodes

FOXNEWS (cable station)   http://www.foxnews.com/   clips and news

*NBC   http://www.nbc.com/   watch episodes

NBCNEWS   http://www.nbcnews.com/   news and clips

Public Telvision

*PBS   http://www.pbs.org/   full free episodes, news

CREATE   http://www.createtv.com/   schedule, episode info, clips, a few episodes

WILDTV   http://www.pbs.org/wnet/wildtv/

**MHZ   http://www.mhznetworks.org/   free full episodes, watch MHZ Worldview free live video streaming

*WORLD   http://worldchannel.org/ free full episodes mostly documentaries

Childrens

PBJ   http://watchpbj.com/ schedule

*PBS Kids http://pbskids.org/ schedule, clips, episodes

QUBO   http://www.qubo.com/home clips, games

*BOUNCETV http://www.bouncetv.com/ Some full episodes of original content, schedule

IONTV http://iontelevision.com/ clips, schedule

MYTV   http://www.mynetworktv.com/   clips

Movies

THIS TV http://thistv.com/ schedule

GET TV http://get.tv/ schedule

MOVIES! http://moviestvnetwork.com/ schedule

ESCAPE TV http://www.escapetv.com/ schedule

Music
**ZUUS COUNTRY http://www.zuus.com/ watch free live TV stream

Living
IONLIFE http://www.ionlife.com/ clips, schedule

*LIVING WELL http://livewellnetwork.com/ watch many free episodes
MYFAMILY TV http://www.myfamilytv.tv/ schedule
Tuff TV http://www.tufftv.com/ full episodes of Auto Wars original programming

Classic TV
RETRO TV   http://www.myretrotv.com/   schedule
*ME TV http://metvnetwork.com/ some free full episodes, schedule
ANTENNA TV http://antennatv.tv/ schedule
COZI TV http://www.cozitv.com/   schedule

Religious
**Atheist TV http://atheists.org/AtheistTV free live TV

**BUDDHIST (Buddhism) http://www.thebuddhist.tv/   free live TV and radio stream

**BYUTV (Mormon)   https://www.byutv.org/ clips, episodes, free live TV
**CBN (Christian)   http://www.cbn.com/   clips, episodes, free live TV
**GODTV (Christian) http://www.god.tv/   episodes, free live TV
**EWTN (Catholic)   http://www.ewtn.com/   news, information, live TV and radio stream

**ISLAM CHANNEL   http://www.islamchannel.tv/   news, information, live TV

**JLTV (Jewish Life)   http://www.jltv.tv/   show info, free live TV
**JN1 (Jewish News)   http://jn1.tv/   news, free live TV
**SHALOMTV (Jewish)   http://www.shalomtv.com/

*means full episodes are available

** means live streaming is available

Paid TV channels (Online Website Links and Resources)

*Has some free unlocked episodes.

**Has live streaming unlocked.

*A&E   http://www.aetv.com/   some full episodes, some locked, clips

ABCFamily   http://abcfamily.go.com/   watch live and episodes with TV provider permission

AHCTV   http://www.ahctv.com/   clips, schedule

AMC   http://www.amctv.com/   full episodes with TV provider permission, extras clips free

Al JEZEERA http://america.aljazeera.com/ clips

*Animal Planet   http://www.animalplanet.com/   some clips, newest finding bigfoot

Anime Network   http://www.theanimenetwork.com/ $6.95/month online

AXS TV http://www.axs.tv/ schedule

BBC America   http://www.bbcamerica.com/   some clips

BET   http://www.bet.com/   show clips and exclusives

*BIO   http://www.biography.com/   some full episodes, full biographies, and mini biographies

**BLOOMBERG   http://www.bloomberg.com/   free live stream USA, Europe, Asia, Event

*BRAVO   http://www.bravotv.com/   clips and some full episodes

**BYUTV (Mormon)   https://www.byutv.org/   clips, episodes, free live TV

**CBN (Christian)   http://www.cbn.com/   clips, episodes, free live TV

ChillerTV   http://www.chillertv.com/ clips, schedule

CLOO   http://www.cloo.com/   clips

*CMTV   http://www.cmt.com/   clips and some full episodes

*CNBC   http://www.cnbc.com/   many full episodes, watch live with TV provider permission

*CNN   http://www.cnn.com/   many clips by show, watch live with TV provider permission

*COMEDY CENTRAL   http://www.thecomedynetwork.ca/   full episodes of main shows

*COOKING   http://www.cookingchanneltv.com/home.html   recipes, watch some episodes

**CSPAN   http://www.c-span.org/   free Live CSPAN, CSPAN2, CSPAN3, CSPAN Radio, also clips

DESTINATION AMERICA   http://www.destinationamerica.com/   clips

*DISCOVERY   http://www.discovery.com/   many Mythbuster and other full episodes, clips

DISCOVERY FIT & HEALTH   http://www.discoveryfitandhealth.com/   clips

*DISNEY   http://disneychannel.disney.com/   some free full episodes, watch live with TV permission

E!   http://www.eonline.com/   clips and news

**ESPN   http://espn.go.com/   clips and news, Watch live ESPN3, Watch other ESPN with TV provider permission

ESQUIRE   http://tv.esquire.com/   watch live and episodes with TV provider permission

*FOX NEWS   http://www.foxnews.com/   clips and news, watch Fox News and Fox Business with TV provider permission

Fox Sports   http://msn.foxsports.com/   clips from shows

*FOOD NETWORK   http://www.foodnetwork.com/   many free full episodes

FUSE   http://www.fuse.tv/   clips

FUSION http://fusion.net/ schedule

*FX   http://www.fxnetworks.com/   clips and some free full episodes

FXX   http://www.fxx.com/   schedule

FXM   http://www.fxnetworks.com/fxm   schedule

G4   http://www.g4tv.com/ clips and news

*Golf Channel   http://www.golfchannel.com/   free full episodes, live with TV provider permission

*GSN (Game Show)   http://gsntv.com/   one of each full free episodes, clips

HALLMARK   http://www.hallmarkchannel.com/   clips

HALLMARK MOVIE   http://www.hallmarkmoviechannel.com/   schedule

*HGTV   http://www.hgtv.com/   many free episodes

*HISTORY   http://www.history.com/   many free full episodes

*H2 (History2)   http://www.history.com/shows/h2   many free full episodes

*HLN   http://www.hlntv.com/   clips and news

**HSN, HSN2   http://www.hsn.com/   all the shopping items, free live TV stream

*IDTV   http://www.investigationdiscovery.com/ full Episodes, clips

*IFC (International Film Channel)   http://www.ifc.com/   they have a free stream room

*LIFETIME   http://www.mylifetime.com/   some free full episodes, some with TV provider permission, some full lifetime movies

*LMN   http://www.mylifetime.com/movies/lifetime-movie-network   a few full episodes

*MLB TV & AUDIO   http://m.mlb.com/network   scores, news, registered user audio, subscriber video stream

*MSNBC   http://www.msnbc.com/   show clips, news, live with TV provider permission

*MTV   http://www.mtv.com/   news, clips, some full episodes, popular music videos

National Geographic   http://channel.nationalgeographic.com/   video clips, locked episodes

National Geographic Wild   http://channel.nationalgeographic.com/wild/   video clips

NBA TV   http://www.nba.com/nbatv/   clips, news, scores, league pass

NFL   http://www.nfl.com/nflnetwork   clips, news, scores, audio and video subscriptions

NHL   http://www.nhl.com/ice/eventhome.htm   clips, news, scores, subscriptions, archives

*NICKELODEON   http://www.nick.com/   many free full videos, games

*NICK JR   http://www.nickjr.com/   activities, games, video clips

*TOONS   http://nicktoons.nick.com/   many free full episodes, clips

OUTDOOR   http://outdoorchannel.com/   clips, info

OXYGEN   http://homepage.oxygen.com/   some clips, episodes with provider permission

OWN   https://www.oprah.com/own   clips

PALLADIA   http://www.palladia.tv/   just schedule

PIVOT TV http://www.pivot.tv/ clips, schedule

*POP TV http://poptv.com/ episodes, schedule

**QVC   http://www.qvc.com/   all the shopping items, free live TV stream

REELZ   http://www.reelz.com/   clips

REVOLT TV http://revolt.tv/ clips

**RT News (Russia Today) http://rt.com/   news, free live TV http://rt.com/on-air/

SCIENCE   http://www.sciencechannel.com/   clips

**SHOPHQ   http://www.shophq.com/   all the shopping items, free live TV streams

*SPIKETV   http://www.spike.com/   many free full episodes

*SMITHSONIAN   http://www.smithsonianchannel.com/   some full episodes

SPEED   http://msn.foxsports.com/speed   clips and news

SUNDANCE   http://www.sundance.tv/   watch with TV provider permission

*SYFY   http://www.syfy.com/   many free full episodes

TBS   http://www.tbs.com/   many full episodes and live with TV provider permission, Ground Floor without permission

TCM   http://www.tcm.com/   watch on demand and live with TV provider permission only

TENNIS   http://www.tennischannel.com/   clips, watch live with TV provider permission

*The Weather Channel   http://www.weather.com/   personal weather info, clips

*TLC   http://www.tlc.com/   a few full original series episode, clips

TNT   http://www.tntdrama.com/   live and episodes with TV provider permission

TruTV   http://www.trutv.com/   clips, episodes with TV provider permission

*CARTOON Network   http://www.cartoonnetwork.com/   some clip, 1 free episode no permission for 10 cartoon series, others require with TV Permission

*TRAVEL   http://www.travelchannel.com/   some free episodes and clips

TVGUIDE NETWORK   http://www.tvgn.tv/   some free full episodes, clips

*TVLAND   http://www.tvland.com/   many free full episodes

*TVONE   http://tvone.tv/   some full episodes

UPTV (formerly Gospel Music Channel)   http://www.uptv.com/music   clips and video

USA   http://www.usanetwork.com/   some free full episodes, most episodes with TV provider permission

VELOCITY   http://www.velocity.com/   clips

*VH1   http://www.vh1.com/   many free full episodes

WWE   http://www.wwe.com/ clips, news, WWE Network with subscription

HBO   http://www.hbo.com/   clips, HBO subscription coming in 2015

CINAMAX   http://www.cinemax.com/   clips

SHOWTIME   http://www.sho.com/   clips

The Movie Channel (owned by Showtime)   http://www.sho.com/site/tmc/videos/home.do

ENCORE   http://www.encoretv.com/   trailers and clips

*STARZ   http://www.starz.com/   free full <u>first</u> episode only of each series

P a g e | **136**

NOTES